Basic Immunogenetics

BASIC IMMUNOGENETICS

Third Edition

H. Hugh Fudenberg
Professor and Chairman
Department of Basic and Clinical
Immunology and Microbiology
Medical University of South Carolina

J. R. L. Pink
Basel Institute for Immunology
Basel, Switzerland

An-Chuan Wang
Professor of Basic and Clinical
Immunology and Microbiology
Medical University of South Carolina

G. B. Ferrara
Director, Pathology and Immunogenetics Laboratory
Institute for the Study and
Treatment of Tumors
Genoa, Italy

New York Oxford
OXFORD UNIVERSITY PRESS
1984

Library of Congress Cataloging in Publication Data
Main entry under title:

Basic immunogenetics.

Bibliography: p.
Includes index.
1. Immunogenetics. I. Fudenberg, H. Hugh.
[DNLM: 1. Immunogenetics. QW 541 B311]
QR184.B37 1984 616.07'9 83-23738
ISBN 0-19-503404-X
ISBN 0-19-503405-8 (pbk.)

Printed in the United States of America

Printing (last digit): 9 8 7 6 5 4 3 2 1

Preface to First Edition

Inherited differences among individuals, or genetic polymorphisms, have frequently been studied by immunological methods. This field of study can loosely be called immunogenetics. Rather than choosing to deal with the whole range of immunogenetics in this text, we have concentrated our attention on areas where the polymorphisms seem to be of particular importance—either because they have been studied in detail (as in the case of some human blood group antigens), because they reveal a wide spectrum of genetic mechanisms operative in producing genetically determined antigens, or because they may raise general questions about individual variations or about regulation of the immune response per se (that is, immune response genes). In this last category are the immunoglobulin polymorphisms, to which we have devoted a large part of the text. Unavoidably, the choice of subjects reflects the writers' own research interests; however, we hope that the examples we have chosen are sufficient to illustrate the variety and potential complexity of other polymorphic systems.

At the end of each chapter is a short bibliography, often containing references to review articles rather than an extensive list of original references. In each chapter subjects are introduced at a level intended not for the specialist but for graduate and medical students with a fair knowledge of biochemistry and genetics. For this reason, many important results have been presented without mention of the authors' names: we apologize in advance to the numerous authors who are not directly acknowledged in the text.

Portions of an article by Dr. J. R. L. Pink that appeared in the *New Scientist* were adapted for use in the Introduction.

H.H.F.
J.R.L.P.
D.P.S.
A.-C.W.

Preface to Third Edition

The third edition of this book has tried to incorporate most of the enormous advances in the field of immunogenetics since the publication of the second edition. Each chapter has been extensively revised in an attempt to cogently present such emerging concepts as immunoglobulin gene rearrangement, evolution, and antibody diversity. Chapter 5 has undergone a major overhaul, reflecting the great increase in knowledge of membrane antigens that occurred subsequent to monoclonal antibody analysis.

A new chapter has been added (Chapter 7) on Immunodeficiency that emphasizes the diversity of the immunological repertoire and the dangers of its dysfunction on homeostasis. Both the glossary of immunologic terms and the blood group terminology have been updated.

It is hoped that this book will serve not only as a reference book for those interested in immunogenetics, but also as an appropriate tool for instructing students, both undergraduate and graduate, in health-related fields.

We are most grateful to Mrs. Linda Paddock and Nancy Butler for superb secretarial assistance. In addition, a special debt of gratitude is owed to Drs. F. Esposito and H. D. Whitten for helping to prepare special crucial sections of the book.

H.H.F.
J.R.L.P.
A.-C.W.
G.B.F.

Contents

Basic Immunogenetics

CHAPTER 1

Introduction

Vertebrates, from primitive fish to humans, possess a well-organized system for defending themselves against infection. When an antigen—a collective term for intruding "foreign" macromolecules such as bacterial coat proteins—enters the bloodstream or lymph, it induces the formation of protective antibody proteins by some of the body's immunocytes. Antibodies are capable of combining with, and usually of precipitating or inactivating, the offending antigen. Figure 1.1 shows antibody molecules (from the high-molecular-weight protein fraction of cow blood) combining with a bacterial flagellum; the antibodies appear rather like staples stuck along the length of the antigen. In the normal course of events, the aggregate formed by the antigen–antibody reaction is removed by the body's scavenging (phagocytic) cells or destroyed by antibodies with the aid of the hemolytic complement system (a mixture of at least nine components, present in the blood, whose destructive action toward bacterial or other cells is triggered by the formation of the antigen–antibody complex).

Antibodies are remarkable for their ability to distinguish antigenic macromolecules from normal body constituents. In addition they can distinguish very subtle differences among the intruding antigens. The high specificity of different antibody preparations was first clearly demonstrated by the German chemist Karl Landsteiner; he found, for example, that rabbit antibodies can distinguish between two antigenic proteins differing only in that a D-tartaric acid residue is attached to one, and its optical isomer, an L-tartaric acid residue, to the other. Figure 1.2 shows some of the compounds used by Landsteiner in his classical examination of the specificity of different antibody preparations.

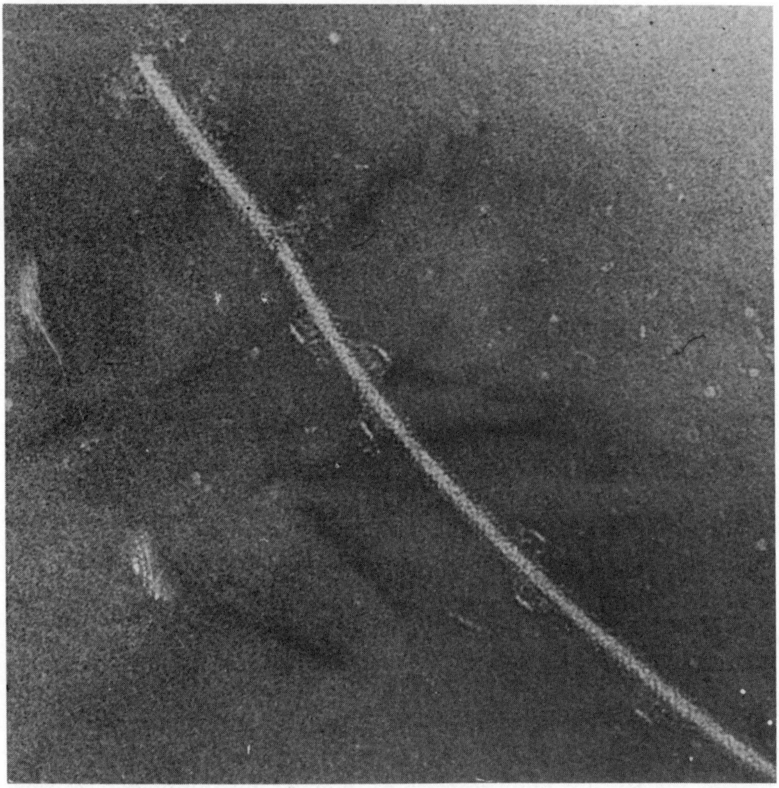

Fig. 1.1 Antibodies from cow serum in combination with a bacterial flagellum. The antibodies appear like staples in this photograph; they are of the IgM class. Magnification × 356,000. (Courtesy of Dr. A. Feinstein)

The differences between these compounds are so slight that the number of different antibodies an individual can make must be very large—perhaps of the order of a million or more.

These considerations raised the first of many puzzles that have intrigued immunologists. The antibodies produced in response to various antigens appeared to differ significantly from each other not in gross physical or chemical characteristics, but only in ability to combine with the different antigens. What makes one antibody different from another? Are the many different antibodies genuinely different proteins (that is, proteins with different amino acid sequences), or are they proteins with identical sequences but with the ability to fold around antigens in a great variety of ways?

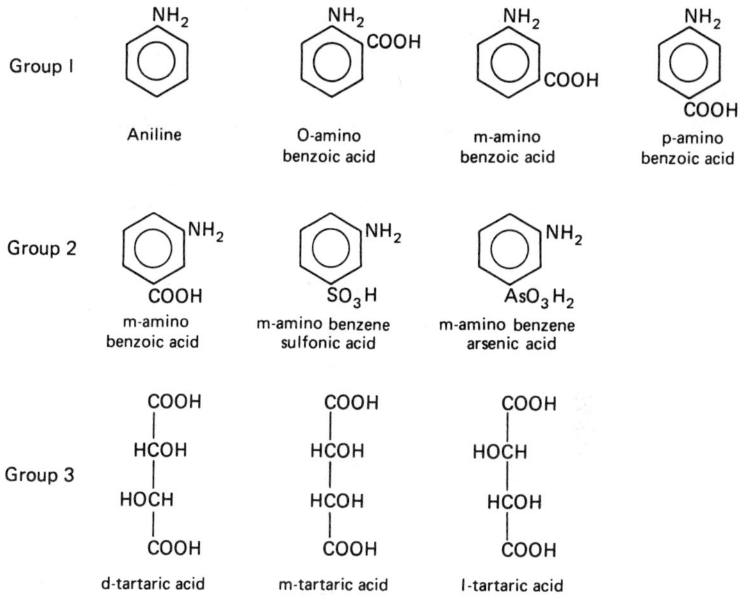

Fig. 1.2 Some groups of compounds used by Landsteiner to demonstrate the high specificity of antibody preparations. By themselves low-molecular-weight compounds are generally incapable of inducing antibody synthesis, but if the compounds are chemically coupled to protein molecules (in which form they are described as haptens) and injected into an experimental animal, antibodies against the hapten (as well as the protein) will be made. Landsteiner observed that antibodies against each hapten (of the groups shown here) were quite distinct, giving no reaction with the other chemically related hapten.

This problem has been resolved: different antibody molecules do have different sequences. However, the question was difficult to answer, for the following reason: almost all antibody preparations examined to date have been heterogeneous. As discussed in Chapter 2, the evolution of the antibody protein system has produced a family of protein chains—of at least 10 distinct types in humans—many or all of which are present in typical antibody preparations. Even within one individual, each type includes an enormous number of variants differing slightly from each other in sequence. These chains can combine with each other in different ways to give variations of the basic antibody protein structure shown in Figure 1.3. The proteins formed in this manner are defined as the immunoglobulins.

Chapter 3 elaborates on a further cause of heterogeneity in anti-

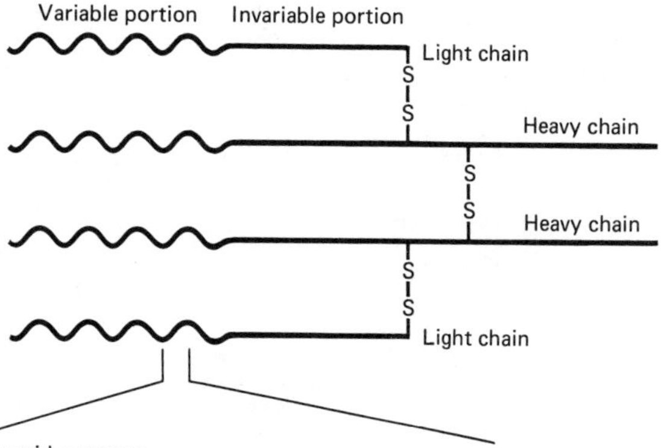

Protein number	Amino acid sequence
1	••• TYR–CYS–GLN–GLN–PHE–ASP–ASN–LEU–PRO–LEU •••
2	••• TYR–CYS–GLN–GLN–TYR–GLU–THR–LEU–PRO–ARG•••
3	••• TYR–CYS–GLN–GLN–TYR–ASP–ASP–LEU–PRO–PRO •••
4	••• TYR–CYS–GLN–GLN–TYR–GLU–ASN–LEU–PRO–TYR •••
5	••• TYR–CYS–GLN–MET–ARG–LEU–GLU–ILE–PRO–TYR •••
6	••• TYR–CYS–GLN–GLN–TYR–GLU–THR–SER–PRO–THR •••

Fig. 1.3 The basic structure of an antibody molecule: four chains (two L and two H) joined by disulfide bridges. The amino-terminal portion of each chain (the V region) differs in different antibodies, while the remainder of the chain (the C region) may be identical in different antibodies.

body preparations: genetic differences (inherited in a Mendelian fashion) may distinguish the immunoglobulins of one individual from those of another. These differences (known as "allotypic" alternatives) have been very helpful in analyzing the genetic basis of antibody variability, since the presence of a genetic marker directly associated with differences in protein structure (in an otherwise homogeneous protein chain) provides good evidence that a single gene (stretch of DNA) codes for that sequence.

These results, although answering previous questions about the structural basis of antibody variability, raise another problem: how is the genetic information necessary to code for perhaps a million different antibodies stored in the body and transmitted from one generation to another? Is there a separate gene for each one of the possible antibody sequences? Is the information stored as single master stretches of DNA that can be altered by mutation during the life of an individual (or perhaps decoded in a number of different ways)? Is the information coded in a few stretches of DNA that recombine among themselves to give many different combinations of sequence? The surprising result of the genetic and structural studies described in Chapters 3 and 4 was that each immunoglobulin chain consists of a variable (V) region, in which the amino acid sequence variability responsible for antibody specificity is located, and a constant (C) region (Fig. 1.3 and Table 1.1). The DNA sequences coding for V and C regions are separate, and a fascinating variety of different mechanisms, discussed in Chapter 4, contributes to generating V-region diversity.

These studies on the genetic control of antibody synthesis naturally raise a further question, discussed in Chapter 5: what are the genetic bases of the many diseases that involve a defect in antibody synthesis? These defects fall into two classes. First, there may be a specific defect in the ability to respond to a particular antigen—for example, some strains of mice or guinea pigs cannot produce an immune response to certain synthetic antigens—a phenomenon subsequently observed in humans with natural allergens. Alternatively, there may be a general lack of immune function—that is, an inability to synthesize any antibodies, or to reject tissue transplants, or both. Such defects, either "antigen-specific" or generalized, are found in certain rare human diseases, most of which clearly have a genetic origin.

One might expect that genes responsible for antigen-specific defects in immune responses would code for immunoglobulin V regions, but in fact only rarely has this correlation been observed. Much more frequently, the genes responsible for antigen-specific immune response defects have been located in a single chromosomal region called the major histocompatibility complex (MHC). This region controls the synthesis of a variety of immunologically relevant proteins, including (1) certain complement components, (2) the major histocompatibility antigens, which are important in defense against viral infections, and (3) the antigens coded by the *Ir* (immune response) genes, to which many antibody response defects have been localized.

In order to understand how the products of these genes function, we need to know in detail how the synthesis of a particular antibody is induced by the presence of the corresponding antigen. Much evidence shows that each antibody-producing cell, with few if any exceptions, makes antibody of only one specificity and, therefore, has been committed to using only one V-region gene for L-chain synthesis and one for H-chain synthesis. Thus individual antibody-synthesizing cells taken from a mouse that has been immunized

Table 1.1 Comparison of the complete amino acid sequences of two human Bence–Jones proteins (from patients Ha and Bo). The sequence of protein Ha is given in full, and only the positions where protein Bo differs in sequence are indicated. The chains are of the λ type, but similar variation between proteins is observed in κ and H chains. Comparison of several λ chains indicates that sequence variation in these chains is observed between positions 1 and 107 (the V region). The remainder of the chain constitutes the C region. PCA is pyrrolidone carboxylic acid (a cyclized glutamate derivative).

1									10										20					
PCA	Ser	Val	Leu	Thr	Gln	Pro	Pro	Ser	Val	Ser	Gly	Thr	Pro	Gly	Gln	Arg	Val	Thr	Ile	Ser	Cys	Ser	Gly	Gly
—	—	Ala	—	—	—	—	—	—	Ala	—	—	Ser	—	—	—	Ser	—	—	—	—	—	Thr	—	Thr
				30										40										50
Ser	Ser.	Asn	Gly	Thr	Gly	Asn	Asn	Tyr	Val	Tyr	Trp	Tyr	Gln	Gln	Leu	Pro	Gly	Thr	Ala	Pro	Lys	Leu	Leu	Ile
—	—	Asp	Val	Gly	Asn	Asp	Lys	—	—	Ser	—	—	—	—	His	—	—	Arg	—	—	—	—	Val	—
									60										70					
Tyr	Arg	Asp	Asp	Lys	Arg	Pro	Ser	Gly	Val	Pro	Asp	Arg	Phe	Ser	Gly	Ser	Lys	Ser	Gly	Thr	Ser	Ala	Ser	Leu
Phe	Glu	Val	Ser	Gln	—	—	—	—	—	—	—	—	—	—	—	—	—	—	Asn	Asp	Thr	—	—	—
				80										90										100
Ala	Ile	Ser	Gly	Leu	Arg	Ser	Glu	Asp	Glu	Ala	His	Tyr	His	Cys	Ala	Ala	Trp	Asp	Tyr	Arg	Leu	Ser	Ala	Val
Thr	Val	—	—	—	—	Ala	—	—	—	—	Asp	—	Tyr	—	Ser	Ser	Tyr	Val	Asp	Asn	Asn	Asn	Phe	?
									110										120					
Val	Phe	Gly	Gly	Gly	Thr	Gln	Leu	Thr	Val	Leu	Arg	Gln	Pro	Lys	Ala	Ala	Pro	Ser	Val	Thr	Leu	Phe	Pro	Pro
—	—	—	—	—	—	Lys	—	—	—	—	—	—	—	—	—	—	—	—	—	—	—	—	—	—
				130										140										150
Ser	Ser	Glu	Glu	Leu	Gln	Ala	Asn	Lys	Ala	Thr	Leu	Val	Cys	Leu	Ile	Ser	Asp	Phe	Tyr	Pro	Gly	Ala	Val	Thr
									160										170					
Val	Ala	Trp	Lys	Ala	Asp	Ser	Ser	Pro	Val	Lys	Ala	Gly	Val	Glu	Thr	Thr	Thr	Pro	Ser	Lys	Gln	Ser	Asn	Asn
				180										190										200
Lys	Tyr	Ala	Ala	Ser	Ser	Tyr	Leu	Ser	Leu	Thr	Pro	Glu	Gln	Trp	Lys	Ser	His	Arg	Ser	Tyr	Ser	Cys	Gln	Val
									210															
Thr	His	Glu	Gly	Ser	Thr	Val	Glu	Lys	Thr	Val	Ala	Pro	Thr	Glu	Cys	Ser								

with two antigens (for example the easily distinguishable red blood cells from sheep and from camels) do not make antibodies to both antigens. In addition the types of antibody-producing cells found in myelomatosis,* which are presumed to be derived from a single parent cell, also produce antibody-like molecules of unique sequence.

These findings are incorporated in a theory of antibody production first advocated by Jerne and by Macfarlane Burnet (the clonal selection theory). According to this theory in its simplest form, antigen can induce antibody synthesis only by contact with a small proportion of cells (antigen-sensitive cells) that have already been committed to making that antibody (Fig. 1.4). The process of commitment of these cells, which takes place during their development and differentiation in the immune system, is still mysterious, but it presumably involves the synthesis of an antibody-like "receptor" molecule that is carried on the cell's surface and that, on contact with antigen, can stimulate the cell to divide and differentiate. The "receptor" has a specificity identical to that of the antibody that the cell and its progeny are destined to secrete. Combination of a receptor with antigen is presumably a trigger that stimulates the antigen-sensitive cell to divide and produce a large number (a "clone") of daughter cells. Cells of a single clone produce antibody molecules of identical structures, but the type of antibody produced by each clone may be different.

These ideas may be used to explain many different observations. However, a variety of experimental results cannot be explained by the clonal selection theory in this simple form. For example, initiation of an immune response, rather than resulting from combination of antigen with single antigen-sensitive cells, in some cases requires the cooperation of more than one cell in recognizing the antigen. It has been shown, on the one hand, that, to provoke an antibody response, such an antigen must have at least two recognizable determi-

* In certain disorders of antibody-producing cells, homogeneous immunoglobulins, for the most part without obvious antibody activity, are produced in the blood of individuals suffering from a malignant proliferation of cells in the bone marrow. In humans, the classical examples of such diseases are (1) Waldenström's macroglobulinemia, characterized by overproduction of homogeneous IgM in the serum and by proliferation of lymphoid cells, and (2) myelomatosis, characterized by overproduction of any of the other immunoglobulin classes (IgG, IgA, etc.) and proliferation of plasma cells with appearance of destructive lesions in bone tissue. In some strains of mice, tumors that secrete immunoglobulins can be induced experimentally (by injecting the mice with certain mineral oils.)

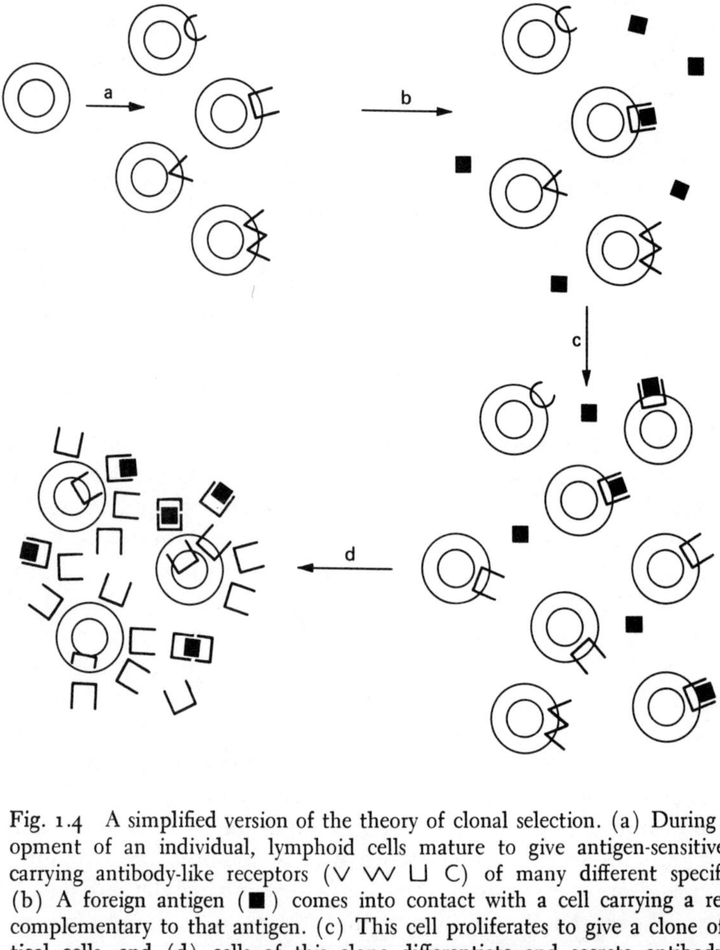

Fig. 1.4 A simplified version of the theory of clonal selection. (a) During development of an individual, lymphoid cells mature to give antigen-sensitive cells carrying antibody-like receptors (V VV ⊔ C) of many different specificities. (b) A foreign antigen (■) comes into contact with a cell carrying a receptor complementary to that antigen. (c) This cell proliferates to give a clone of identical cells, and (d) cells of this clone differentiate and secrete antibody of a single specificity. If the antigen is encountered a second time on a later occasion, more antigen-sensitive cells are available to proliferate, and these will produce an increased (secondary or anamnestic) immune response to that antigen.

nants on its surface, and, on the other hand, that antigen recognition leading to antibody production (at least for certain antigens) probably requires the interaction of at least two separate lymphoid cell populations. One population (probably derived directly from the bone marrow) is responsible for antibody production, and the other undergoes differentiation in the thymus and is at least partly responsible for antigen recognition and for some specific immunological functions (such as graft rejection) whose exact mechanisms are still unknown. This duality is paralleled by the empirical division of immunological reactions into those mediated by antibody molecules present in the serum or in secretions (humoral immunity) and those —including graft rejection—that are mediated by cells but that, because of their specificity, are thought to depend on the presence of antibody-like molecules on the surfaces of these cells (cellular immunity). Chapter 5 deals with the possibilities of using these results to account for genetic aberrations in antibody synthesis, and with the role of the major histocompatibility complex in cell–cell interactions.

Many of the findings we have mentioned depended on the use of concepts or techniques borrowed from fields not directly related to immunology. For example, the techniques used to determine the amino acid sequence of an immunoglobulin molecule and the concepts that helped relate this structure to the function and evolution of the molecule (Chapter 2) were, for the most part, developed in the course of work on insulin, hemoglobin, ribonuclease, and other proteins investigated by molecular biologists. The mapping of the immunoglobulin gene linkage groups (Chapter 3) has involved using the techniques of population genetics, serology, protein chemistry and molecular biology. These studies could not have been carried out without the results from clinical and chemical laboratories leading to characterization of myelomatosis and isolation of the proteins produced in the course of the disease.

Several other techniques have recently resulted in crucial advances in immunogenetics, including assortive breeding leading to the development of congenic, coisogenic, and recombinant inbred strains of experimental animals (Chapter 5); nucleic acid hybridization, whereby enzymatically cleaved strands of nucleic acid are allowed to react with a radiolabeled DNA or RNA probe fragment (Chapter 4); nucleic acid sequencing, whereby individual nucleic acid bases may be sequenced after modification and cleavage on polyacrylamide gel electrophoresis; recombinant DNA cloning, the process whereby an isolated gene is inserted into a vector plasmid or bacteriophage prior to transformation or infection of a bacterial host; and gene mapping by endonuclease cleavage—a very precise process in which cleavage of DNA segments by a series of different, specific endonucleases allows the determination of the order of the gene map.

Immunology is already repaying this debt owed to many other disciplines. The concepts (1) that two or more genes may code for a single polypeptide chain, (2) that families of multiple, closely linked, closely related genes may be basic units of vertebrate genomes, and (3) that gene rearrangement may occur during differentiation are now central to eukaryotic molecular biology. The techniques of radioimmunoassay, tissue typing, and monoclonal antibody production are widely used in many fields of both basic and clinical research. The further development of immunology will, we hope, continue to prove useful in many biological fields and in the provision of new methods for eradication, prevention, diagnosis, or treatment of human disease.

RECOMMENDED READINGS

Books

Klein, J. (1982), *Immunology, The Science of Self–Nonself Discrimination* (1st ed.). New York: Wiley, pp. 156–255.

Wachtel, S. S. (1982), *H-Y Antigen and the Biology of Sex Determination.* New York: Grune & Stratton.

Bailey, D. W. (1982), How pure are inbred strains of mice? *Immunol. Today* 3, 210.

Articles

Williams, A. F. and Gagnon, J. (1982), Neuronal cell Thy-1 glycoprotein: homology with immunoglobulin. *Science* 216, 696.

CHAPTER 2

Immunoglobulin Structure and Evolution

Almost 40 years ago, Tiselius demonstrated that antibody activity was present mainly in the electrophoretically slowly moving (more positively charged) protein fraction isolated from the serum of immunized animals. Later improvements in the technology of protein fractionation led to isolation of the most obvious serum immunoglobulin components: these are IgG, which is present in large amounts, even in normal sera, and IgM, which has an unusually high molecular weight (almost 1 million).

The most effective procedures for purification of these molecules involve precipitation by high salt or alcohol concentrations followed by electrophoretic separation, or by chromatography on an ion-exchange (diethylaminoethyl cellulose) column. The high molecular weight of IgM makes its additional fractionation by gel filtration or sucrose gradient centrifugation an effective purification procedure. Isolation of some immunoglobulin classes (IgM and IgA) from normal serum has been accomplished by a useful general method—the use of an antiserum (anti-IgM or anti-IgA) that has been rendered insoluble by a chemical cross-linking reaction. IgM or IgA can be removed from serum by using this insolubilized antibody preparation and later eluted from the antibody by washing with an acidic buffer. A specific method such as this is required for the purification of the minor immunoglobulin classes IgD and IgE, since their concentration in normal serum is so low.

It became clear that immunoglobulins from normal sera were very heterogeneous in antigenic characteristics, in electrophoretic mobility, and even in amino acid sequence. A similar degree of heterogeneity was evident in most specific antibody preparations; in addition most of these preparations were clearly mixtures of anti-

Table 2.1 Human immunoglobulins and their constituent chains (adapted from Nisonoff et al., 1975)

	Chain	Approx. mol. wt. × 10^{-3}	No. of domains	Approx. % CHO	Corresponding class or subclass	Approx. conc. in serum, mg/ml
HEAVY	$\gamma 1$	50	4	2–3	$G1 = \gamma 1_2 L_2$	9
	$\gamma 2$	50	4	2–3	$G2 = \gamma 2_2 L_2$	3
	$\gamma 3$	60	4	2–3	$G3 = \gamma 3_2 L_2$	1
	$\gamma 4$	50	4	2–3	$G4 = \gamma 4_2 L_2$	0.5
	$\alpha 1$	55	4	7–11	$A1 = \alpha 1_2 L_2$	3
	$\alpha 2$	50	4	7–11	$A2 = \alpha 2_2 L_2$	0.5
	μ	70	5	9–12	$M = (\mu_2 L_2)_5 J$	1.5
	δ	60	4	9–14	$D = \delta_2 L_2$	0.03
	ϵ	70	5	12	$E = \epsilon_2 L_2$	0.0003
LIGHT	κ	25	2	–	–	–
	λ	25	2	–	–	–
OTHER	J chain	15	–	7	–	–
	Secretory component	70	–	20	–	–

bodies with different affinities for antigen. Attempts to fractionate or purify these antibodies until they appeared homogeneous or had a unique binding constant for antigen were consistently unsuccessful.

An important development was the realization that the homogeneous myeloma and Bence–Jones proteins were related to normal immunoglobulins. The availability of these proteins in large quantities and in a homogeneous form permitted an unambiguous classification and nomenclature of human immunoglobulins and their constituent chains to be established. Studies of these proteins also showed that, in general, each immunoglobulin chain, in its amino-terminal half (of an L chain) or quarter [of a heavy (H) chain], had a unique sequence. (Discussion of the variability of immunoglobulin chains is deferred to Chapter 4.) In contrast the remainder of the chain (the constant or C region of the chain) is identical in immunoglobulin molecules of a given class and subclass (apart from the occurrence of genetically polymorphic forms of the different classes and subclasses; see Chapter 3). Thus the C region defines each particular class and subclass of immunoglobulin chain, and, since each C-region subclass is antigenically distinct, these can be determined immunologically.

IMMUNOGLOBULIN H AND L CHAINS

The major classes of human immunoglobulins are IgG, IgA, and IgM. Two additional classes (which make up less than 1% of the total body immunoglobulins), IgD and IgE, have been identified in human serum. These classes (IgG, A, M, D, and E) are defined by the presence of specific H chains (γ, α, μ, δ, and ϵ chains) (Table 2.1). Two L-chain types (κ and λ) have been defined, and each class of H chain may combine with either κ or λ chains. (The heavy–light chain combination is not completely random; thus there are slight but statistically significant differences between the proportions of IgG, A, and M proteins with either κ or λ chains.)

These major H- and L-chain types, although clearly distinct from each other, appear to be derived from a common ancestor early in the evolution of the immune system, as is discussed later in this chapter. Thus there is a clear amino acid sequence homology when the structures of human γ, μ, and α chains, or of κ and λ chains, are compared (Table 2.2). More recent evolution by gene duplication of some of these chain types has occurred; thus in humans four

Table 2.2 Amino acid sequence homology of human H- and L-chain types placed by Ser and Lys, respectively.

Region being compared	Chain	Sequence
Carboxy-terminal sequence (around disulfide bridge to H chain)	κ	Val Thr Lys Ser Phe Asn Arg Gly Glu Cys
	λ	Val Glu Lys Thr Val Ala Pro Thr Glu Cys Ser
Carboxy-terminal sequence	α	Met Ala Glu Val Asp Gly Thr Cys Tyr
	μ	Met Ser Asx Thr Ala Gly Thr Cys Tyr
Sequence around disulfide bridge to L chain	μ	Pro Leu Val Ser Cys Glx Asx
	γ*	Pro Leu Ala Pro Cys Ser Arg
	δ	Pro Ile Ile Ser Gly Cys Arg
	ϵ	Pro Leu Thr Arg Cys Cys Lys

* Gamma chains of subclasses $\gamma 2$, $\gamma 3$, $\gamma 4$; in $\gamma 1$ chains Cys and Arg are replaced by Ser and Lys respectively.

closely related γ-chain subclasses (γ1, γ2, γ3, and γ4 chains) have been defined, each with a different sequence and different biological properties. The sequence homology between subclasses is around 90%. Each subclass of chains is synthesized under the control of a different gene, although these genes are closely linked (Chapter 3) and all subclasses are present in any normal individual. Subclasses of γ chains have been defined in many mammalian species (p. 36).

Two subclasses of human α chains, distinguishable by their antigenic and chemical characteristics, have been described. The homology between the α1 and α2 chains is as great as that between the γ-chain subclasses. The existence of μ- and δ-chain subclasses has not been excluded in species other than mouse, where DNA cloning and sequencing experiments indicate their absence. Light-chain subtypes* (isotypes) seem less common than γ-chain subclasses, at least for κ chains; however, two λ-chain isotypes have been described in guinea pigs, three in mice, and at least four in humans. The major mouse isotypes, λ1 and λ2, are not very closely related (about 70% sequence identity between C and 90% between V regions), whereas the human isotypes differ only slightly from one another (Table 2.3).

The most likely explanation for these findings is that the λ-chain isotypes are coded for by separate genes, although other possibilities are not ruled out: for example, an unusual DNA triplet coding for residue 188 in a single human λ-chain gene might allow insertion of either lysine or arginine at this position by a (specific) transfer RNA molecule. This explanation seems unlikely, because analogous cases of duplicated, closely related genes are known in other systems. Thus the γ chains of human fetal hemoglobin are normally a mix-

Table 2.3 Amino acid substitutions in four human λ-chain isotypes. Isotype numbering is arbitrary.

λ-Chain type	Position no.				
	111	113	157	167	188
λ1	Ala	Ser	Ser	Thr	Arg
λ2	Ala	Ser	Ser	Thr	Lys
λ3	Ala	Ser	Gly	Thr	Arg
λ4	Asn	Thr	Gly	Lys	Arg

* The term *types* is currently used to distinguish κ and λ chains, and *classes* to distinguish λ, μ, α, δ, and ε chains.

ture of two almost identical chains that differ in the substitution of a glycine for an alanine residue at position 136. Here it can be verified that two different genes encode the two chains, since occasional mutants in one gene product do not affect the other.

THE IMMUNOGLOBULIN FOUR-CHAIN UNIT

In 1959 Porter used the proteolytic enzyme papain to split IgG (isolated from rabbit serum) into fragments that retained biological activity, and Edelman showed that the IgG molecule could be separated into its component H and L chains after reduction with mercaptoethanol. Porter proposed a four-chain model of the IgG molecule and assigned (approximately) the point where the papain split it. His model (Fig. 2.1) has been very little altered since its proposal. The molecule consists of identical pairs of H and L chains

Fig. 2.1 The four-chain, basic structural unit of an IgG molecule (CNBr, cyanogen bromide). Homology units are labeled.

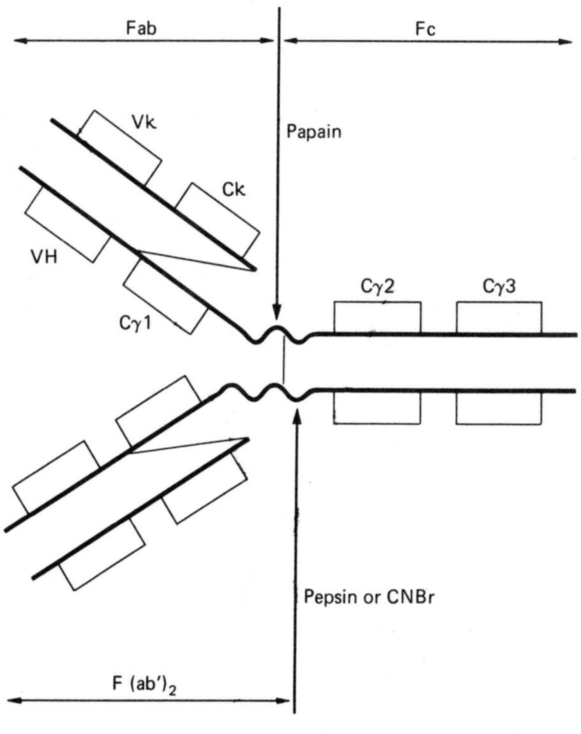

(of molecular weights around 50,000 and 25,000 respectively) connected by disulfide bridges. Papain digestion produces two Fab (for antigen-binding) fragments and one Fc (for crystallizable) fragment. Each Fab fragment contains one L chain and one Fd fragment (approximately half of an H chain) and is capable of binding one antigenic determinant (a complete IgG molecule has two antigen-binding sites).

One extension of this model has been the finding that different IgG subclasses, or IgG molecules from different species, have different numbers of interheavy-chain disulfide bridges (Fig. 2.2). The

Fig. 2.2 Distributions of interchain disulfide bridges (thin lines) joining L and H chains (short and long thick lines) in different immunoglobulin classes and subclasses. The positions of the bridges are taken by comparison with homologous sequences in human IgG1 molecules. In some cases it is not ruled out that the marked bridges are labile intrachain rather than interchain bridges. Human IgG3 probably has 11 interchain bridges.

The exact number of inter-H chain disulfide bridges (bracketed) in IgA molecules is uncertain, as are the positions in the H chain of the bridges in mouse IgA and IgE. The bridge involved in covalent dimerization of IgA molecules has not been established. Am$^+$ and Am$^-$ are allotypic forms of human IgA2 molecules.

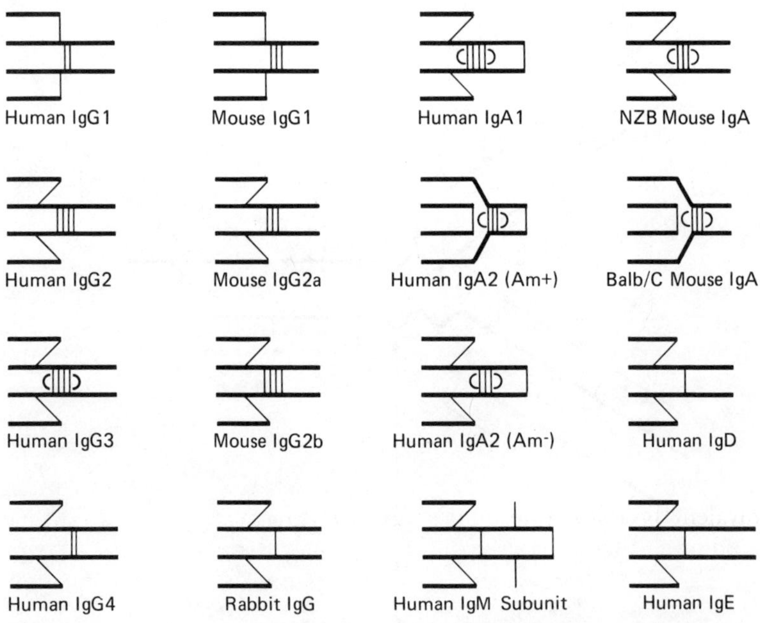

region near the middle of the H chain where these bridges are located is very sensitive to enzymic digestion, not only by papain, but by pepsin, trypsin, and other enzymes. The products of digestion vary in different cases. The usual products are an Fc-like fragment and an Fab-like fragment, or an $F(ab)_2$-like disulfide-bridged dimer form if the point of splitting is to the carboxy-terminal side of the interheavy-chain bridges (for example, after pepsin or cyanogen bromide digestion). After papain digestion the Fc fragment consists of a dimer of two H-chain fragments (each of molecular weight of about 20,000) held together by noncovalent interactions. Since a low concentration of a reducing agent such as mercaptoethanol is necessary to activate the enzyme for digestion, the interheavy-chain disulfide bridges may not be left intact after treatment with papain. Pepsin digestion normally cleaves the Fc portion of the chain into smaller peptide fragments.

Electron micrographs of IgG molecules in contact with antigen show a Y-shaped molecule that has a degree of flexibility about the center of the Y: this point is presumably the location of the interchain bridges, since pepsin digestion (which removes the Fc fragment of the molecule) leaves visible only the arms of the Y, which therefore correspond to an $F(ab)_2$-like dimer. The site of antigen binding is at the tips of these arms.

The four-chain structural unit typical of an IgG molecule is also the basis of the structure of other immunoglobulins. The minor immunoglobulin classes IgD and IgE exist as simple four-chain units, of molecular weights about 180,000 and 200,000, respectively (the ϵ H chain of the IgE molecule has a molecular weight of about 70,000). Molecules of the IgD class are extremely susceptible to digestion by proteolytic enzymes, which degrade the IgD to small peptide fragments. Fab- and Fc-like fragments can be isolated after pepsin digestion of IgE.

An IgM molecule in serum contains a total of five four-chain units, joined by disulfide bridges (Fig. 2.3). The molecule can be broken down into units by very mild reduction of only the interunit bonds. Stronger reduction conditions (0.1 M mercaptoethanol) will also produce four-chain units, but in these the chains are held together only by noncovalent interactions.

Since the IgM molecule in serum is essentially a pentamer of the divalent IgG structure, it is surprising that the valency of IgM (in binding antigen) has occasionally been measured as 5 rather than 10. Fab-like fragments can be made from IgM by limited enzyme

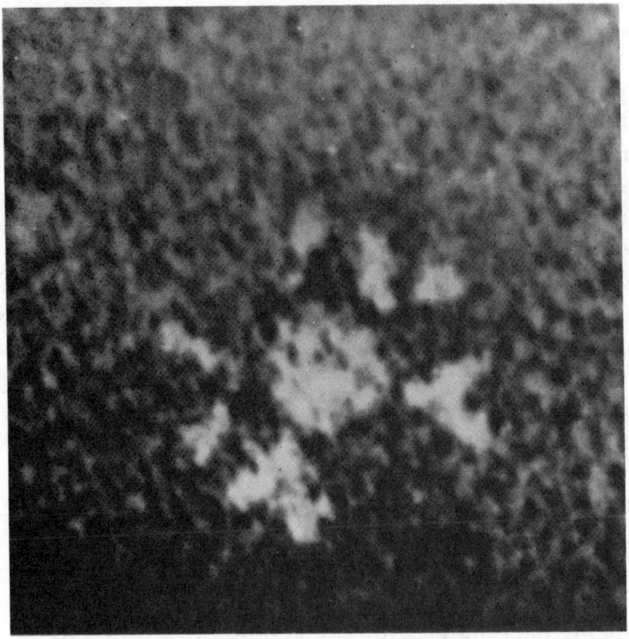

Fig. 2.3 An IgM molecule (from dogfish serum). The molecule is basically a circular polymer of five Y-shaped units; the bases of the Y form a central disk, to which the arms of the Y (corresponding to Fab fragments) are flexibly attached. The disk's diameter is about 160 Angstroms. An edge-on view of the molecule can be seen in Figure 1.1. (Courtesy of Dr. A. Feinstein)

digestion; if these fragments are made from a pentavalent IgM molecule, each fragment contains one binding site, a fact suggesting that some binding sites may be either conformationally altered or sterically blocked by large antigens in the complete molecule.

The avidity of IgM antibodies for polyvalent antigens (as measured by, for example, the ability of measured amounts of IgM antibodies to agglutinate red blood cells) is usually much higher than the avidity of corresponding IgG antibodies. This is thought to result from the availability of one IgM molecule of many antigen-binding sites (rather than any intrinsic properties of the binding sites themselves). A certain degree of flexibility of the IgM subunits in the whole molecule is required for this argument. Electron micrography of IgM molecules (Fig. 2.3) clearly shows that the five four-chain units are linked in a circle and that the links are probably flexible, so that the units may bend above and below the plane of the circle (see also Fig. 1.1).

IgA is found not only in serum but also in body secretions (such as saliva, nasal secretions, sweat, and breast milk). In serum the most usual form of IgA consists of one four-chain unit (whose molecular weight is approximately 160,000), but this unit may polymerize to give disulfide-bonded polymers with 8-chain, 12-chain, or larger structures. The IgA found in secretions consists of two four-chain units associated with a third chain type—the secretory component. The molecular weight of the whole complex is about 400,000. The secretory component is a single peptide chain, with a molecular weight of approximately 70,000, that is bonded to a secretory IgA dimer by strong noncovalent interactions and also (to most human IgA molecules) by disulfide bonds. Its function may be to enable IgA antibodies to be transported across body membranes into secretions. It is not made by the plasma cells that synthesize the body's IgA but rather by nonmotile cells near the points of the membrane where secretion occurs. There does not seem to be any close structural relationship between the secretory component and any immunoglobulin chain, and in fact the component can be found free in agammaglobulinemic patients who lack measurable IgA in their serum or secretions.

In addition, almost all IgA polymers and IgM proteins (from cases of human IgA myeloma or Waldenström's macroglobulinemia) contain an extra chain (J chain or "joining" chain) that is distinct from the secretory component and from H and L chains in electrophoretic mobility, amino acid composition, and molecular weight (about 15,000). The stoichiometry of the J-chain immunoglobulin complex is thought to be one J chain per polymer; in the case of IgM the J chain is disulfide-bonded to the H chains of two of the IgM subunits. It seems likely that J chain is required for efficient intracellular polymer formation (although *in vitro* polymers can in some cases be formed in the absence of J chain).

IMMUNOGLOBULIN DISULFIDE BRIDGES AND DOMAINS

Figure 2.2 shows the ways in which (as far as is known) the *inter*-chain disulfide bridges are arranged in different immunoglobulins. In most immunoglobulin molecules these bridges may be reduced selectively (leaving the intrachain bridges intact) if the molecule is reduced in a nondenaturing solvent, so that the conformation of the individual chains is largely maintained. Figure 2.2 and Table 2.4 illustrate two unusual aspects of these bridges: (1) their great variety, in number and even in position, in different immunoglobulin classes and subclasses and (2) the unusual nature of the amino acid sequences around these bridges. The number of interheavy-chain bridges may vary from one to as many as four or five in different molecules. Usually these bridges are concentrated in a region in the middle of the chain, and often this area contains an exceptional concentration of proline residues. This part of the H chain has been called the "hinge" region, because a study of electron micrographs of IgG molecules suggests that the molecules are flexible at the point of susceptibility to proteolytic digestion by pepsin.

An interheavy–light-chain bridge is formed between H chains and the carboxy-terminal residue of κ L chains, or the residue adjacent to the carboxy-terminus of λ chains. In the H chain the point of attachment varies—from near the center of the chain (in human and mouse IgG1 molecules) to a position about 100 residues removed toward the amino-terminus of the chain in most other immunoglobulins (Fig. 2.2). In some IgA molecules this bridge is formed between two L chains, and only noncovalent interactions keep the H and L chains together. (Interlight-chain bonding is also found in Bence–Jones proteins, which are commonly secreted as disulfide-bridged dimers.)

The *intra*chain disulfide bridges of immunoglobulin chains are spaced regularly along each chain (Fig. 2.1), two in an L chain and four in a γ chain. Each bridge encloses a loop of about 60 to 70 residues. This pattern seems to be invariant except that an additional bridge may sometimes be present, in either a C region (for example, in rabbit γ chains) or in a variable (V) region. Examination of the amino acid sequences around the cysteine residues of each bridge reveals some clear homologies between the sequences, and the analysis of more nearly complete sequence data shows that the homologies extend over the region of the entire disulfide-bridged loop. Each of these homology regions or domains is about 110 residues long. There is weak but definite homology between V and C regions, a circumstance suggesting that any evolutionary divergence of C and V regions from a common ancestor was a very early event. However, there is an obvious homology between all of the C-region domains. Table 2.5 shows the homologies between C regions of the L and H

Table 2.4 Amino acid sequence around the interheavy-chain disulfide bridges in several IgG molecules

Species	IgG subclass																		
Human	IgG1	Lys	Val	Glu	Pro	Lys	Ser	Cys*	Asp	Lys	Thr	His	Thr	Cys	Pro	Pro	Cys		
	2	Thr	Val	Glx	Arg	Lys	Cys	Cys	Val	Glx	□	□	□	Cys	Pro	Pro	Cys		
	3	Cys	Pro	Glu	Pro	Lys	Ser	Cys	Asp	Thr	Pro	Pro	Pro	Cys	Pro	Arg	Cys		
	4	Arg	Val	Glu	Ser	Lys	Tyr	Gly	Pro	Pro	□	□	□	Cys	Pro	Pro	Cys		
Rabbit	IgG	Thr	Val	Ala	Pro	Ser	Thr	Cys	Ser	Lys	Pro	□	Thr	Cys	Pro	Pro	Pro		
Mouse	IgG1								Asp	Cys*	Gly	Cys	Lys	Pro	Cys	Ile	Cys		
	IgG2a							Gly	Pro	Thr	Ile	Lys	Pro	Cys	Cys	Pro	Pro		
	IgG2b										Ile	Asn	Pro	Cys	Pro	Pro	Cys		
Guinea pig	IgG2	Thr	Val	Glu	Pro	Ile	Arg	Thr	Pro	Glx	Pro	Asx	Pro	Cys	Thr	Cys	Pro	Lys	Cys
Goat	IgG							Lys	Cys	Pro	Lys	Glx	Pro	Cys	Val	Arg	□		

□ = Deletion or insertion introduced to maintain sequence homology.
* = Disulfide-bridged to light chains.

Table 2.5 Comparison of amino acid sequences of different homology regions in a single IgG molecule (protein Eu). (For nomenclature of regions, see Figure 2.1. Deletions, indicated by dashes, have been introduced to maximize homologies.)

							110									120								
Eu C_L (Residues 109-214)						Thr	Val	Ala	Ala	Pro	Ser	Val	Phe	Ile	Phe	Pro	Pro	Ser	Asp	Glu	Gln	—	—	Leu
Eu C_{H1} (Residues 119-220)						Ser	Thr	Lys	Gly	Pro	Ser	Val	Phe	Pro	Leu	Ala	Pro	Ser	Ser	Lys	Ser	—	—	Thr
Eu C_{H2} (Residues 234-341)						Leu	Leu	Gly	Gly	Pro	Ser	Val	Phe	Leu	Phe	Pro	Pro	Lys	Pro	Lys	Asp	Thr	Leu	Met
Eu C_{H3} (Residues 342-446)						Gln	Pro	Arg	Glu	Pro	Gln	Val	Tyr	Thr	Leu	Pro	Pro	Ser	Arg	Glu	Glu	—	—	Met
				130										140										
Lys	Ser	Gly	Thr	Ala	Ser	Val	Val	Cys	Leu	Leu	Asn	Asn	Phe	Tyr	Pro	Arg	Glu	Ala	Lys	Val	—	—	Gln	Trp
Ser	Gly	Gly	Thr	Ala	Ala	Leu	Gly	Cys	Leu	Val	Lys	Asp	Tyr	Phe	Pro	Glu	Pro	Val	Thr	Val	—	—	Ser	Trp
Ile	Ser	Arg	Thr	Pro	Glu	Val	Thr	Cys	Val	Val	Val	Asp	Val	Ser	His	Glu	Asp	Pro	Gln	Val	Lys	Phe	Asn	Trp
Thr	Lys	Asn	Gln	Val	Ser	Leu	Thr	Cys	Leu	Val	Lys	Gly	Phe	Tyr	Pro	Ser	Asp	Ile	Ala	Val	—	—	Glu	Trp
	150										160										170			
Lys	Val	Asp	Asn	Ala	Leu	Gln	Ser	Gly	Asn	Ser	Gln	Glu	Ser	Val	Thr	Glu	Gln	Asp	Ser	Lys	Asp	Ser	Thr	Tyr
Asn	Ser	—	Gly	Ala	Leu	Thr	Ser	Gly	—	Val	His	Thr	Phe	Pro	Ala	Val	Leu	Gln	Ser	—	Ser	Gly	Leu	Tyr
Tyr	Val	Asp	Gly	—	Val	Gln	Val	His	Asn	Ala	Lys	Thr	Lys	Pro	Arg	Glu	Gln	Gly	Tyr	—	Asp	Ser	Thr	Tyr
Glu	Ser	Asn	Asp	—	Gly	Glu	Pro	Glu	Asn	Tyr	Lys	Thr	Thr	Pro	Pro	Val	Leu	Asp	Ser	—	Asp	Gly	Ser	Phe
						180										190								
Ser	Leu	Ser	Ser	Thr	Leu	Thr	Leu	Ser	Lys	Ala	Asp	Tyr	Glu	Lys	His	Lys	Val	Tyr	Ala	Cys	Glu	Val	Thr	His
Ser	Leu	Ser	Ser	Val	Val	Thr	Val	Pro	Ser	Ser	Ser	Leu	Gly	Thr	Gln	—	Thr	Tyr	Ile	Cys	Asn	Val	Asn	His
Arg	Val	Val	Ser	Val	Leu	Thr	Val	Leu	His	Gln	Asn	Trp	Leu	Asp	Gly	Lys	Glu	Tyr	Lys	Cys	Lys	Val	Ser	Asn
Phe	Leu	Tyr	Ser	Lys	Leu	Thr	Val	Asp	Lys	Ser	Arg	Trp	Gln	Gln	Gly	Asn	Val	Phe	Ser	Cys	Ser	Val	Met	His
	200													210										
Gln	Gly	Ley	Ser	Ser	Pro	Val	Thr	—	Lys	Ser	Phe	—	—	Asn	Arg	Gly	Glu	Cys						
Lys	Pro	Ser	Asn	Thr	Lys	Val	—	Asp	Lys	Arg	Val	—	—	Gly	Pro	Lys	Ser	Cys						
Lys	Ala	Leu	Pro	Ala	Pro	Ile	—	Glu	Lys	Thr	Ile	Ser	Lys	Ala	Lys	Gly								
Glu	Ala	Leu	His	Asn	His	Tyr	Thr	Gln	Lys	Ser	Leu	Ser	Leu	Ser	Pro	Gly								

chains from protein Eu, a human IgG1, κ myeloma protein studied in detail by Edelman. When the L-chain C (Cκ) region is compared with the three portions of the H-chain C region (the domains called Cγ1, Cγ2, and Cγ3), about 30 to 35% of the residues compared are identical in each case. The homology between domains is equally striking when the three-dimensional structures (as determined by X-ray crystallographic methods—see Chapter 4) of, for example, the L- and H-chain C-region domains of an Fab fragment are compared (p. 119).

It seems clear from these studies that all H chains have evolved by duplication of "homology region" genes, each coding for approximately 100 to 110 amino acid residues. In accordance with this idea, studies of amino acid sequence have shown that the μ and ε chains have one such unit (corresponding to an increase in molecular weight of about 12,000) more than the well-studied γ chains.

Biosynthesis of Immunoglobulins

In plasma cells, each of which is engaged in the synthesis and secretion of immunoglobulin molecules at a rate of about 2000 molecules per second, immunoglobulin chains are synthesized on membrane-bound ribosomal aggregates (polysomes) carrying messenger RNA (mRNA) molecules. Polysomes that carry mRNA coding for γ heavy chains contain about 16 to 20 ribosomes, those coding for light chains, 7 to 8 ribosomes. The techniques used for isolating immunoglobulin-chain mRNA from mouse plasmacytomas involve extraction of RNA from membrane-bound polysomes and its further purification by sucrose gradient centrifugation and electrophoresis in acrylamide gels. A further purification step makes use of the fact that the mRNA contains a stretch of about 200 adenine residues (added after transcription of the RNA from DNA) and therefore can be selectively retained on columns to which the complementary oligonucleotide (poly-*d*-thymidine) has been bound.

The mouse light-chain mRNA is about 1200 nucleotides long. Of these, about 320 code for the C and 320 for the V region, and most of the rest are untranslated (Fig. 2.4). However, the direct product of translation of mouse light-chain mRNA in a cell-free system is a precursor molecule, longer by more than 20 amino acids than the secreted light chain. Similar leader sequences have been identified on newly synthesized heavy chains. The leader sequences are removed during or immediately after synthesis of the complete Ig polypep-

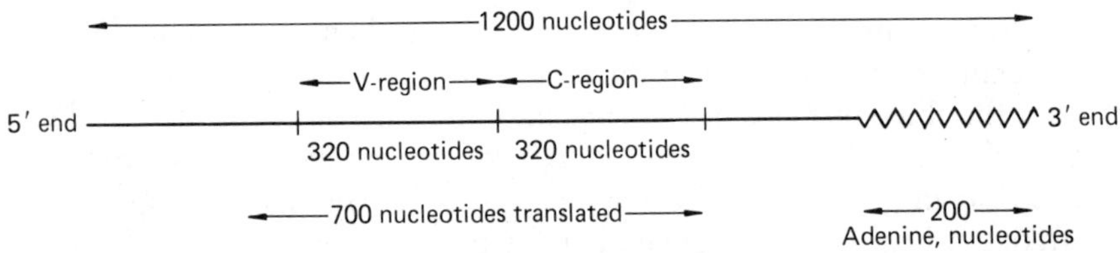

Fig. 2.4 Schematic representation of L-chain mRNA.

tide, and are probably responsible for ensuring that the chain is synthesized on membrane-bound polysomes before secretion.

Additional intermediates in the biosynthesis of immunoglobulins have been identified by pulse-labeling techniques in cultures of myeloma cells. The "two heavy–two light" chain structure of an IgG immunoglobulin (abbreviated as H_2L_2) can be built up in several ways by using either L_2, H_2, or HL units. Usually when synthesis of a single myeloma immunoglobulin is investigated, it is found that one or the other of these intermediates predominates, although some proteins may use more than one route of assembly. The major determining factor seems to be the strength of interaction between C regions of the chains involved, for proteins of the same H-chain subclass tend to be assembled in similar ways.

Polymerization of IgM and IgA four-chain units to give the final polymeric structures takes place immediately before release of the molecules from antibody-secreting cells and, as noted above, involves the incorporation of J chain into the secreted polymer.

All normal immunoglobulin heavy chains are glycosylated (Fig. 2.5). The carbohydrate chains are of two types: simple di- or trisaccharides linked to the hydroxyl group of threonine or serine residues, and complex oligosaccharides linked to the asparagine amido nitrogen. These latter chains are formed in two stages: first, a mannose-rich core sugar chain is transferred to the protein from a glycolipid donor, and this chain is subsequently modified by removal of some mannose residues and addition of terminal sugars (mainly N-acetyl glucosamine, galactose, and sialic acid). The initial steps in glycosylation occur while the growing polypeptide is still attached to the polysome, but intermediate forms lacking terminal sugars can be detected in suitable pulse-chase experiments.

In many small lymphocytes, immunoglobulin molecules are present on the cell membrane. Most bone marrow-derived (B) lympho-

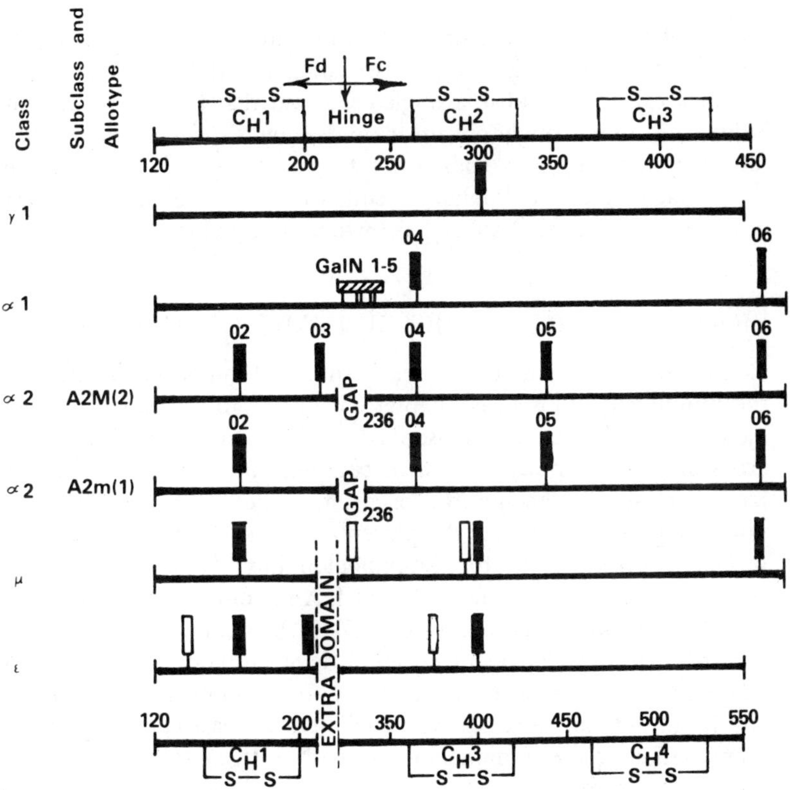

Fig. 2.5 Location of carbohydrate in human γ1, α1, α2, μ, and ε heavy chains. The vertical rectangles signify glucosamine oligosaccharides. Shading indicates those that have homologous positions in two or more chains. The numbers in the upper scale give the residue positions for the γ and α chains; those in the lower scale, for the μ and ε chains. To facilitate comparison, the extra domain (Cμ2 and Cε2, respectively) has been omitted in the μ and ε chains. An oligosaccharide is present in Cε2, but not in Cμ2. The symbols 02–06 and GalN1–5 denote α-chain oligosaccharides. (From A. Toraño *et al.*, *Proc. Natl. Acad. Sci.* (*USA*) *74*: 2301, 1977, with permission)

cytes carry both IgM and IgD at their surface, but other isotypes are also present on about 3 to 5% of B cells. Membrane IgM, unlike secretory IgM, is a single four-chain H_2L_2 unit. This unit is not simply a subunit of the secretory pentamer; its μ chains have, at their C-terminal end, a 41-residue hydrophobic amino acid sequence instead of a 20-residue hydrophilic region present in secretory chains.

Studies of the biosynthesis of these two chain types have shown that the two chains are synthesized on two different mRNA species (both of which can be present in a single cell), but that these mRNAs can be derived from a single precursor mRNA by a splicing process that can remove nucleotides coding for the secretory tail, at the same time attaching the hydrophobic sequence-coding region to the rest of the μ sequence. RNA splicing is further discussed in Chapter 3.

BIOLOGICAL PROPERTIES OF IMMUNOGLOBULINS

Combination with antigens is clearly the most important function of the immunoglobulins (see Chapter 4), but a large number of other biological properties can be ascribed to these molecules. Such properties are, for example, fixation of complement, binding to the surface of mast cells, monocytes, or macrophages, ability to be transported (specifically) across the placenta, and ability to combine with proteins specific for secretory immunoglobulins (Table 2.6). Some of these functions, such as the ability of IgA to combine with the secretory component or the ability of IgG to be transported across the placenta, are not dependent on an antigen–antibody contact. Other properties, however, such as the ability to fix complement or to give rise to an inflammatory reaction mediated by mast cells, depend on combination of the immunoglobulin with antigen—or strictly speaking (in the case of complement fixation by IgG) perhaps only on aggregation of several immunoglobulin molecules: chemically aggregated IgG molecules can fix complement in the absence of any con-

Table 2.6 Biological properties of different human immunoglobulins

Property	IgG1	G2	G3	G4	IgM	IgA1,2	IgD	IgE
Complement fixation	+	+	+	−	+	−	−	
Passage across placenta	+	±	+	+	−	−		−
Association with "secretory component"	−	−	−	−	−	+	−	−
Binding to human mast cells (reaginic activity)	−	−	−	−	−	−	−	+
Anaphylactic reaction with guinea pig skin	+	−	+	+	−	−	−	

tact with antigen, and, in general, complement is not fixed by antibody in combination with a monovalent hapten.

It has been suggested that each domain has evolved to meet one or more specific biological needs of the molecule's C regions. The binding sites for complement (or rather the first component of complement, which can associate directly with immunoglobulin molecules) are located in the Cγ2 region of γ chains, since fragments have been isolated (by limited enzymic digestion or cyanogen bromide cleavage) that can still fix complement and that derive from the Cγ2 portion of the molecule. The Cγ3 domain is probably the part of the molecule responsible for maintaining the immunoglobulin at the cell membrane in the case of those IgG molecules that are present at the lymphocyte surface or are bound to monocytes, since this domain is not available for reaction with antisera at the cell membrane. The immunoglobulins found at the surface of the majority of lymphocytes are IgM and IgD; in the case of IgM the molecule exists in the membrane as a single four-chain subunit (molecular weight 180,000) rather than as the much larger pentamer; however, it is not clear what factors are responsible for the fact that the molecule remains in a lipid-rich environment (the membrane) instead of being secreted into the surrounding medium.

The major immunoglobulin classes are clearly distinguishable in biological properties. IgM is the first immunoglobulin to be synthesized by the fetus and the first antibody class to be produced in a normal response to injection of a novel antigen into blood or lymph. Most or all IgM molecules fix complement, an action leading to cell lysis (if the IgM antibody is directed against a cell-surface antigen) with a very high efficiency. (Approximately 1000 times as many IgG as IgM molecules are needed to induce lysis of a red cell. This is thought to be because one IgM molecule can fix complement, whereas two or more IgG molecules in contact are necessary for fixation.)

Why then do we synthesize IgG at all? One property associated only with IgG molecules is the ability to provide protection for the fetus and newborn by IgG transferred across the placenta, but this seems an insufficient reason for the major importance of IgG in all species examined. A more general explanation has been proposed based on the observation that an individual cell may change from synthesizing IgM to IgG antibody during an immune response (see Chapter 3) and on the findings that passive infusion of IgG antibody into an animal will produce an inhibition of synthesis of that antibody. Infusion of IgM has no such marked effect (and may perhaps even stimulate antibody production). The role of a "switch" from IgM to IgG production may therefore be to inhibit overproduction of antibody during an immune response.

From this point of view it is interesting that certain disease states may involve continuous production of antibody to the patient's own components. How the synthesis of these antibodies is induced is not clear. Of the IgM class the most common antibodies of this type are rheumatoid factors and the cold agglutinins (see Chapter 4, p. 105). Perhaps a failure of a switch mechanism can account for their continued production over periods documented as long as 10 years.

The biological role of IgD is not known. Its concentration in serum is very low (usually less than 0.1 mg/ml), but remarkably, IgD is present (often together with IgM) on the surface of a majority of human immunoglobulin-bearing lymphocytes (Chapter 3). In cells bearing both IgD and IgM the specificity for binding antigen of the two immunoglobulins is the same. In spite of this it has been very difficult to show that IgD antibodies in serum can be induced by antigenic challenge. Thus both IgD and IgM seem to function as lymphocyte receptors for antigen, but the mechanism by which the lymphocyte is stimulated to produce antibody (by foreign antigens) or inactivated (by self-antigens) is still not understood and is highly controversial.

The various subclasses of human IgG molecules have different biological properties (Table 2.6); the same is true of subclasses in other species. Human IgG1 and IgG3 molecules fix complement, IgG2 does so poorly, and IgG4 molecules not at all. (Isolated, heat- or chemically aggregated myeloma proteins can be used to demonstrate complement fixation.) IgG2 molecules are relatively deficient in the ability to be actively transported across the placenta and to bind to monocytes and macrophages. (These cells have on their surfaces receptors specific for IgG1 and IgG3 molecules; the receptors presumably enable the cells to carry out their function of ingesting and destroying antigen complexed with antibody.)

IgA clearly has a distinct function: its predominance in human body secretions has already been mentioned. A large part of the body's IgA is synthesized near the sites of its secretion by a system of plasma cells that is essentially independent of the IgM- and IgG-producing system, both in physical location and in function (although the precursors of all plasma cells are derived from the bone marrow). Thus local antigenic stimulation of a secretory organ may induce immunoglobulin production but not raise serum IgM or IgG antibody levels or markedly alter the small level of circulating IgA antibodies, and, vice versa, serum antibody levels may rise during infections without evidence of corresponding changes in levels of secretory IgA.

The existence of a separate immunoglobulin class responsible for allergic reactions was predicted by Ishizaka, who was able to remove (using a specific antiserum) the activity in serum responsible for these reactions without removing IgG, M, or A. The class has been named IgE, and the finding of several myeloma proteins of the IgE type has fully confirmed its existence. The interaction of a mast cell with IgE–antigen complexes induces the cell to release a number of pharmacologically active substances (for example, histamine) that give rise to the symptoms of allergic reactions. The normal function of this cycle of events is not clear but may be to induce a local inflammatory reaction that would be advantageous in, for example, dealing with parasitic infection.

Finally some specific antibody activities may be preferentially associated with immunoglobulins of a particular class or subclass. It is not surprising that the route of administration of an antigen should influence the class of antibody formed, since, as already mentioned, IgA is the predominant antibody in (human) secretions, whereas serum immunoglobulins are predominantly IgG. It is, however, more surprising that certain antigens may preferentially stimulate production of a single IgG subclass. Thus immunization of human volunteers with dextran usually results in production of antibody of the IgG2 subclass. Antibodies to a specific blood-clotting factor (Factor VIII) in several cases of a rare form of hemophilia have been reported to be mainly IgG4 immunoglobulins. These findings are unexpected (and unexplained), for (see Chapter 3) the antigen-binding sites associated with each immunoglobulin class and subclass are probably identical or very similar.

IMMUNOGLOBULINS IN DIFFERENT SPECIES

All vertebrates probably have a well-developed immune response. The ability to give an immune response has been investigated in many mammalian species, in several species of birds, and in a number of reptiles, amphibians, and fish. The lower orders of vertebrates investigated include the elasmobranchii (sharks and rays) and the even more primitive cyclostomata (lampreys and hagfish). A specific antibody response could be induced in each species examined, and in all cases this response was mediated by serum proteins that resembled the well-described mammalian immunoglobulins. Thus the immune system evolved in the ancestors of today's vertebrates at least 300 million years ago.

Several invertebrates have been examined for the ability to give an immune response. The species examined in some detail include the lobster, the horseshoe crab (*Limulus*), earthworms, starfish, and several insect species. Many of these species contain, in their body fluids, substances that agglutinate red cells of other species, sometimes in a rather specific manner, but the molecules responsible for this agglutinating activity do not resemble mammalian immunoglobulins in molecular weight, polypeptide chain structure, or stability (in the few cases where they have been examined). The main difficulty in this type of study is to decide which criteria define the ability to give an immune response. These criteria might include, for example, ability to give a secondary (increased and more rapid) response, high specificity of the response, and execution of the response by cells and molecules with at least some similarity to those involved in the immune reactions of higher species. Thus far no immune response meeting all these criteria has been shown to occur in any invertebrate, although in most cases sufficient information to judge the third criterion is not available and is difficult to obtain because of the small size and feeble responses of the animals involved.

However, the lamprey and the even more primitive hagfish (which are both chordates and therefore derived from a line immediately ancestral to the vertebrates) provide a specific response after repeated stimulation with bacterial antigen or red cells. The molecules with antigen-binding activity are present in very low amounts but clearly resemble the immunoglobulins of higher organisms. Thus, in the lamprey, antibody activity was localized in two proteins having sedimentation coefficients of 7S and 14S and appearing to contain H and L chains similar in molecular weight to the immunoglobulin chains of higher organisms. The 14S component may have been a polymerized form of the 7S component. The molecules' H and L chains were not joined by interchain disulfide bonds. Thus the earliest immunoglobulin-like molecules to appear in evolution are found in the most primitive organisms to possess an organ resembling a thymus.

Several species of sharks have been studied for their capacity to produce antibodies. Typical findings in four species (dogfish and nurse, lemon, and leopard sharks) are that two types of immunoglobulins can be distinguished: a low-molecular-weight form (with sedimentation coefficient 7S) and a high-molecular-weight form sedimenting at 17 or 19S. Both forms are probably based on a four-chain unit, for both give rise to H and L chains of molecular weights approximately 70,000 and 22,000, respectively. In all cases the 19S (or 17S) form had properties suggesting that it was a simple polymer of the 7S form; in some cases the peptide maps, and in other cases the antigenic characteristics, of the two forms were identical. Thus these primitive immunoglobulins have apparently not evolved to give the classes (IgM, IgA, IgG) typically found in mammalian species; rather, they seem to be derived from a single precursor of these classes. It is interesting that this precursor in certain respects (ability to polymerize, molecular weight of the heavy chain) resembles mammalian IgM rather than IgA or IgG immunoglobulins.

Rather similar findings have been reported in the few studies of fish immunoglobulins that have been made. Diversification of the immunoglobulin's basic structure to give rise to distinct H-chain classes seems to have begun at this stage, for two species of lungfish (a species rather closely allied to the line along which the vertebrates evolved) have been shown to synthesize (besides 19S and 7S immunoglobulins) a 5S immunoglobulin characterized by an H chain of low molecular weight (35,000). This unusual class of chain has been named a ν chain and has no obvious counterpart in mammalian species; however, an H chain of similar molecular weight has been isolated from a low-molecular-weight immunoglobulin found in ducks. An unusual IgM-like immunoglobulin, comprising four four-chain units instead of five, has been isolated from two species of fish (carp and grouper), but it is not yet certain that the H chains of these molecules are distinct from the H chains of low-molecular-weight fish immunoglobulins.

The first clear evidence of separate classes corresponding to mammalian IgM and IgG in nonmammalian species was obtained from bullfrogs, whose 19S and 7S antibodies clearly possess H chains derived from the 19S and 7S proteins (molecular weights 72,000 and 53,000) and resembling those of mammalian μ and γ chains. Whether amphibians make any immunoglobulin resembling IgA is unknown; cells resembling plasma cells have been found in the gut of amphibians (toads) but not in fish. It is not clear whether these cells are synthesizing IgA, but it is possible that the requirement for a secretory immunoglobulin became important with the development of land-living species.

The two well-studied avian species (chicken and ducks) synthesize 19S and 7S immunoglobulins probably homologous to IgM and IgG. However, the H chains from chicken "IgG" have a higher molecular weight than mammalian γ chains, so that the relationship between avian and mammalian 7S immunoglobulins is still unclear. A secretory immunoglobulin homologous to mammalian IgA has been isolated from chicken bile. In addition the duck immunoglobulins have been reported to include a 5.7S species with an H chain of relatively low molecular weight (possibly homologous to the lungfish ν chain). The first evidence that two types of L chains (homologous to the mammalian κ and λ) underwent evolutionary divergence in nonmammalian species came from the study of L chains from chickens. These L chains have an amino-terminal sequence resembling that of human λ rather than κ chains. Many mammals appear to have both κ and λ chains (on the basis of the similarity of sequence of the carboxy-terminal region of the chain to human κ or λ chains). In contrast, most of the sequence data (derived from the study of the amino-terminal portions of L chains) from lower species, such as sharks, are ambiguous or insufficient for the assignment of the chains to the κ and λ class. A clear estimate of when the κ/λ divergence took place is impossible to obtain from the phylogenetic studies available at present, but since the degree of sequence similarity between κ and λ is slightly greater than that between μ and γ, it might be expected that the κ/λ divergence took place after the μ/γ divergence of amphibians from the mammalian line (see above).

The immunoglobulins of humans, mice, and rats have been particularly well studied because of the availability of myeloma proteins in these species. However, several other species, representative of all the main mammalian classes, have been examined. In general IgM- and IgG-like molecules have been readily identified in all these species, and subclasses of IgG (which are difficult to define in the absence of data obtained from myeloma proteins) have been present in every species that has been carefully studied. IgA-like molecules have been identified in a large number of mammalian species and probably exist in all mammals. One difficulty in these studies has been that in certain species (for example, cattle and sheep), IgD-like molecules are also present in secretions such as colostrum in quantities sufficient to prevent the easy characterization of IgA. In these species newborn animals receive most of their maternal antibody in their mother's colostrum rather than by transfer across the placenta, so that secretion of IgG as well as IgA may be a necessity for protection of the newborn.

The other two immunoglobulin classes, IgD and IgE, are present in very low levels in human sera and presumably in sera of other species. Nevertheless the characteristic biological property of IgE, the ability to sensitize cells to give an allergic reaction, has permitted its identification in a number of species. Ideally the identification of IgE should rest on additional criteria such as cross-reaction with human IgE, clear lack of cross-reaction with γ and α chains, and relatively high molecular weight (human IgE, 200,000); by most or all of these criteria IgE has been identified in monkeys, rabbits, mice, and dogs. IgE myeloma proteins have also been obtained from a strain of rats susceptible to spontaneous myelomatosis.

The lack of easily definable biological properties in human IgD makes it difficult to identify IgD in other species. However, the relatively large degree of divergence of IgD from IgG in terms of amino acid sequence and antigenic properties suggests that most or all mammalian species may possess IgD-like immunoglobulins. An immunoglobulin cross-reacting with human IgD has been identified in various primates, and an IgD-bearing lymphoid cell line derived from mice has been described.

Most or all mammalian species have at least two IgG subclasses, which differ in biological properties, electrophoretic mobility, and other properties (Table 2.7). The criteria used to subdivide IgG into subclasses in several species have been (1) clear-cut differences between electrophoretic mobilities of different subclasses, (2) antigenic differences, and (3) differences in biological properties. It is clear from amino acid sequence studies that IgG subclasses have evolved independently in several species. Thus the γ chains of three human subclasses have the carboxy-terminal sequence of *Leu*-Ser-*Leu*-Ser-*Pro*-Gly (Table 2.8), whereas the γ chains of the two bovine subclasses have the homologous carboxy-terminal sequence *Thr*-Ser-*Lys*-Ser-*Ala*-Gly. This occurrence within one species of very similar subclasses could have two explanations: either there have been independent doublings of the γ-chain genes in different species, or there has been convergent evolution of the γ-chain genes in each species. Because there is much evidence (for example from studies of genes coding for hemoglobin chains in several species) that gene duplication is a common process in vertebrate evolution, and because the selective

Table 2.7 IgG subclasses in various species

Species	IgG subclasses	Relative electrophoretic mobility	Complement fixation
Human	IgG1	Slow	+
	2	Slow	+
	3	Slow	+
	4	Fast	−
Mouse	IgG1 (IgF)	Fast	−
	2a(IgG)	Slow	+
	2b(IgH)	Slow	+
	3	Slow	
Guinea pig	IgG1	Fast	
	2	Slow	+
Cow	IgG1	Fast	+
	2	Slow	±
Horse	IgG (T)	Fast	−
	IgG (a)	Slow	+
	IgG (b)	Slow	+
	IgG (c)	Slow	−
Rat	IgG1	Fast	
	2a,b	Slow	
	2c	Very slow	
Sheep	IgG1	Fast	+
	2	Slow	±
Goat	IgG1	Fast	
	2	Slow	
Dog	IgG a, b, c	Slow	
	d	Fast	

pressures needed to produce convergent evolution of γ-chain sequences would be extraordinarily high, the first explanation appears more satisfactory.

Gamma-chain subclasses having a common evolutionary origin may, however, be shared by closely related species. Thus apes apparently have IgG H chains similar at least to the human $\gamma 2$ and $\gamma 3$ chains (since immunological cross-reaction between the human subclasses and gorilla and chimpanzee IgG fractions has been observed). The mouse $\gamma 1$ and $\gamma 2b$ subclasses differ more from each other in se-

Table 2.8 C-terminal sequences of H chains from various species*

Chain	Sequence																	
Human γ1	-His-	Glu-	Ala-	Leu-	His-	Asn-	His-	Tyr-	Thr-	Gln-	Lys-	Ser-	Leu-	Ser-	Leu-	Ser-	Pro-	Gly-
Human γ2	—	—	—	—	—	—	—	—	—	—	—	—	—	—	—	—	—	—
Human γ3(Gm)	—	—	—	—	—	—	Arg-	Phe	—	—	—	—	—	—	—	—	—	—
Human γ3(Gm)	—	—	—	—	—	—	Arg-	—	—	—	—	—	—	—	—	—	—	—
Human γ4	—	—	—	—	—	—	—	—	—	—	—	—	—	—	—	—	Leu	—
Bovine γ1	—	—	—	—	—	( )	—	—	—	—	—	—	Thr	—	Lys	—	Ala	—
Bovine γ2	—	—	—	—	—	( )	—	—	Met	—	—	—	Thr	—	Lys	—	Ala	—
Sheep γ1	—	( )	—	—	—	( )	—	—	—	( )	—	—	Ile	—	Lys-	Pro	—	—
Sheep γ2	—	( )	—	—	—	( )	—	—	—	( )	—	—	Ile	—	Lys-	Pro	—	—
Horse γT	—	—	—	Val-	Glu	—	—	—	—	—	—	Asn-	Val	—	His	—	—	—
Horse γ(slow)	—	—	—	—	—	—	—	—	—	—	—	—	Val	—	Lys	—	—	—
Dog γa, b	—	—	—	—	—	—	—	—	—	—	—	—	—	—	His	—	—	—
Dog γd	—	—	—	—	Gln	—	—	—	—	Asp-	Leu	—	—	—	His	—	—	—
Guinea pig γ1	—	—	—	—	Pro	—	—	Val-	Ile	—	—	—	Ile	—	Arg	—	—	—
Guinea pig γ2	—	—	—	—	—	—	—	Val	—	—	—	—	Ile	—	Arg	—	—	—
Rabbit γ	—	—	—	—	—	—	—	—	—	—	—	—	Ile	—	Arg	—	—	—
Mouse γ1	—	—	Gly	—	—	—	—	His	—	Glu	—	—	—	—	His	—	—	—
Mouse γ2a	—	Gln-	Gly	—	—	—	—	Val-	Ser-	Thr	—	—	Phe	—	Arg-	Thr	—	—

* Solid lines indicate sequence is same as human γ1 sequence. Amino groups of bracketed residues have not been assigned.

quence than the human γ chain subclasses do, suggesting that counterparts of the mouse γ chains exist in a wide range of rodent and perhaps other mammalian species. Similarly the goat and sheep γ-chain subclasses seem to have had a common evolutionary origin, again based on the criterion of immunological cross-reaction. However, the efficiency of this method of detecting subclasses depends heavily on the specificity of the antisera used, and the evolution of IgG subclasses is best studied by comparison of amino acid sequence data.

In some species IgG subclasses that have independent evolutionary origins have evolved rather similar biological properties (Table 2.7). Thus electrophoretically fast-moving IgG subclasses present in several species (termed IgG4 in humans but IgG1 in guinea pigs) mediate passive cutaneous anaphylaxis reactions (in guinea pig skin) but do not fix complement well. However, the evolutionary significance of these findings is reduced by the fact that in other species (for example, cattle) it is the slow-moving subclass that fixes complement less well. In addition the classical tests for complement fixation and passive cutaneous anaphylaxis reactions involve nonphysiological conditions: often the fixation of guinea pig complement and the anaphylactic reaction of guinea pig skin are investigated rather than the interactions of IgG with homologous complement or skin. Until such tests are done under approximately physiological conditions the significance of some of the subclass-specific differences in biological properties given in Table 2.8 must be considered doubtful.

MECHANISMS OF IMMUNOGLOBULIN EVOLUTION

Since its emergence the immunoglobulin system has evolved to give a complex family of protein chains. Three main stages in this evolution may be summarized as (1) formation of V and C regions, (2) formation of H and L chains, producing a functional immunoglobulin molecule; (3) formation of the major classes of H and L chains, and (4) the more recent evolution of subclasses and subtypes of many of these chains. Point mutation and gene duplication have been fundamental processes leading to diversification at each stage. A scheme for evolution of immunoglobulin chains (Fig. 2.6) may be drawn up by using the results of phylogenetic and amino acid sequence studies. In doing this, use is made of the fact that the rate of accumulation of point mutations in a given protein type seems to remain fairly constant throughout evolutionary time. Thus

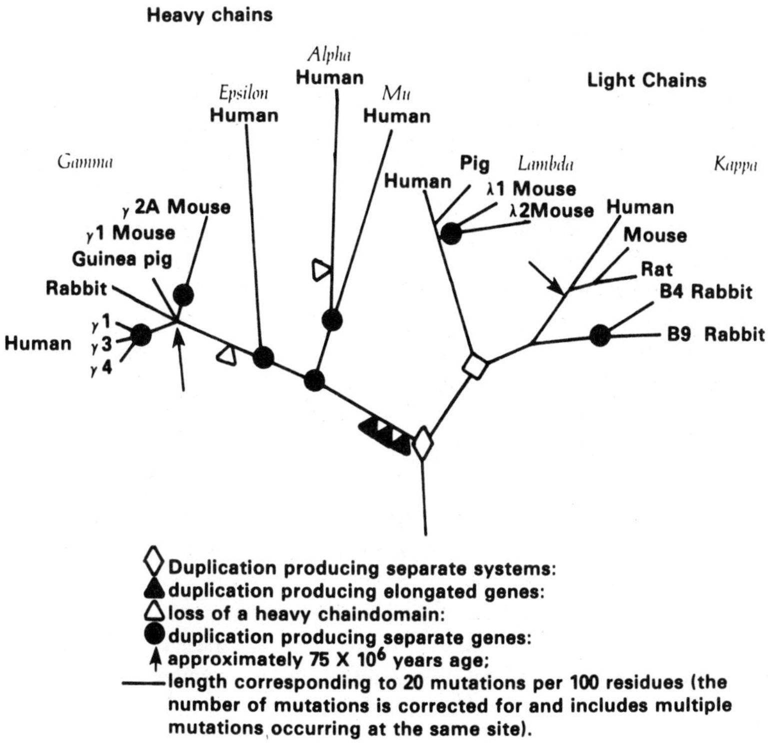

Fig. 2.6 Evolutionary tree of sequenced IgC regions (Barker *et al.*, 1978). Branches leading to γ and λ chains are dashed because there are additional solutions almost as good as those shown. The early divergence of rabbit from other κ chains is unexpected, as is the large difference between allelic B4 and B9 rabbit κ sequences; the two allelic forms may be derived from two different κ genes that had duplicated at an early stage (see text).

the degree of similarity between two homologous proteins is a measure of how recently they evolved from a common ancestral protein.

The V region of an L or H chain bears a certain resemblance to the molecule's C-region homology units. This resemblance is mainly in the length of the regions (about 110 residues) and in the size and position of the disulfide-bridge loop of each region (Table 2.5). A similarity slightly greater than would be expected from random between sequences of V and C regions has also been found. These similarities, although by no means conclusive, suggest that V- and C-region genes are derived from a common ancestral gene. If such a separation of V- and C-region genes did occur, it must have taken place very early in the development of the immune system; for most purposes the evolution of immunoglobulin C regions seems to have been quite separate from that of the V regions. The little that is known about V-region evolution is discussed in Chapter 4.

The C regions of H and L chains are clearly related to each other. The sequence homology between L- (human κ and λ) and H- (human γ) chain homology regions is illustrated in Table 2.5 and in all cases is about 30 to 40%. The homology between the individual γ-chain domains is also about 30 to 35%. The simplest explanation for these similarities includes two stages: first, duplication of a gene coding for an ancestral C region produced L- and H-chain C-region precursor genes. Since present-day L- and H-chain genes are gentically unlinked and probably on different chromosomes, the original gene duplication may have resulted from a process of chromosome gain or tetraploidization. Later, tandem duplications of the primitive H-chain gene gave rise to a larger gene coding for a peptide chain at least twice as long as an L chain. In fact, since primitive vertebrates synthesize μ-like H chains whose C regions are probably four times as long as that of an L chain, at least two gene duplications are necessary to account for the evolution of an H-chain C region from a primitive "L-chain C-region" gene. The emergence of IgG and IgM as separate classes probably took place before the divergence of amphibians and mammals.

The sequence data available from studies of human α, δ, and ε chains suggest that these classes diverged soon after the evolution of separate μ- and γ-chain classes. Information from sequences of different Ig chains in other species on the emergence of chain types and classes is scant, but the rather similar degrees of sequence homology found between γ and μ, and between κ and λ chains, suggest that the γ/μ and κ/λ duplications took place at about the same time.

Recent direct studies of immunoglobulin light- and heavy-chain genes have added much to our knowledge of their evolution. As discussed in Chapter 3, these experiments fully confirmed the long-suspected separation of V- and C-region genes, and, in addition showed that, for each heavy chain, separate DNA regions code for separate domains and, where they are present, for hinge and tail regions as well. Thus, for example, four separate coding sequences (three for the three Cγ domains and one for the hinge) are present in the mouse Cγ1 gene.

As expected from the amino acid sequence data, there is strong (about 60%) homology between nucleotide sequences of corresponding domains of different subclasses and a weaker (about 40%) homology between different domains of the same chain. Surprisingly, one 17-nucleotide region of the mouse γ2b $C_{H}1$ and $C_{H}2$ domains can be matched almost perfectly in an out-of-phase reading frame, so that the corresponding amino acid sequences are quite different; this suggests that a frameshift mutation has occurred, and been tolerated, during the evolution of the domains. The sequences at the borders of the intervening DNA also show homology to each other; otherwise, the intervening sequences in γ1 and γ2b chains are less homologous than are the coding sequences. A similar rapid divergence of noncoding DNA sequences has been noted in comparisons of other eukayotic genes, for example, rabbit and mouse globin genes.

Intervening sequences of similar lengths separating domain-size coding regions exist in μ-, δ-, and α-, as well as γ-chain genes, suggesting that the primitive heavy-chain gene, ancestral to μ, γ, δ, ε, and α, already contained noncoding sequences. Thus, during evolution of heavy-chain classes and subclasses, it is theoretically possible that domains have been exchanged by unequal crossing over between related genes. However, evolutionary trees prepared from comparisons of individual domains are consistent with trees derived from comparisons of entire protein sequences of human heavy-chain C regions, with the exception of the $C_{H}1$ domain for which an alternative, slightly more parsimonious tree can be constructed. Whether this exception is due to domain exchange during evolution is unknown.

Such sequence comparisons also show that the different heavy-chain domains have evolved at different rates: for example, human μ and α chains are 50% homologous in their carboxy-terminal domains, but only 20% homologous in the adjacent domain. When human α, γ, μ, and ε chains were compared, the $C_{H}1$ domains showed the most divergence; the rapid divergence of $C_{H}1$ domains was also observed in a comparison of human, murine, and canine μ chains—but not in a comparison of mouse γ1, γ2a, and γ2b chains, whose $C_{H}1$ domains were 75–80% homologous, as against 55–60% homology for the $C_{H}3$ domains. How these different rates are related to the functions of different domains is unclear.

The structure of the heavy-chain genes also suggests mechanisms for the evolution of hinge regions, and for the production of common heavy-chain variants. In the mouse γ2b gene, there is a 38% homology in nucleotide sequence between the $C_{H}1$ domain and a region of the longest intervening sequence (that preceding the hinge) plus the hinge itself. This intervening region, plus the hinge, therefore may be derived from what was originally an extra domain, corresponding to that present in μ chains (which, in line with this idea, lack a hinge region). Heavy-chain protein variants include deletions of V or C domains, deletions of the hinge region, and production of hybrid γ chains; they are further discussed in the next chapter.

A very rough time scale for immunoglobulin evolution can be arrived at by assuming a constant rate of accumulation of mutational differences between two genes that once had a common ancestor (that is, the number of differences is approximately proportional to the time since their divergence from the ancestral gene). This assumption is approximately true for the evolution of hemoglobins and cytochromes, where sufficient amino acid sequences are available to allow correlations between divergence of sequences and evidence of common ancestors (from the dating of fossil "ancestral species") to be made. Each protein type has a characteristic rate of accumulation of mutations (presumably because of different selective pressures to remain constant); thus, at least in vertebrates, the approximate rates for hemoglobin chains (about 145 residues long) and cytochrome c (about 100 residues long) are 0.14 and 0.02 residue per chain per million years. The rather limited data for mouse, rabbit, and human immunoglobulins (only C regions being considered) suggest a faster rate, about 0.2 mutant per 100-residue C region every million years (Table 2.9).

Table 2.9 "Evolutionary rates" of different proteins.
Expressed as number of amino acid differences out of total number of residues compared

Proteins compared	Amino acid differences
κ chain common region: mouse/human	43/110
Hemoglobin α chain: mouse/human	17/141
Hemoglobin β chain: mouse/human	23/145
γ chain Fc region: rabbit/human $\gamma 4$	38/120
Hemoglobin α chain: rabbit/human	25/141
Hemoglobin β chain: rabbit/human	14/145
Cytochrome c: rabbit/human	9/104

The figures indicate that the rate of amino acid substitution during evolution of κ C-region and Fc region has been roughly twice as fast as in hemoglobin chains.

This estimate of the rate fits well with other evidence: for example, one can estimate (from comparison of sequences of γ chains) that the human γ-chain subclasses evolved from a common ancestor about 20 to 30 million years ago (since they are more than 90% similar to each other). At about this period the hominid line (apes and humans) was probably separating from the lines leading to baboons, rhesus monkeys, and others, and this is consistent with the occurrence of antigenic specificities typical of human subclasses on ape γ globulins, but not on monkey γ globulins. Similarly the degree of identity (less than 40%) between human μ and γ chains suggests their derivation from a common ancestor about 200 million years ago: this is consistent with the findings of separate μ-like and γ-like chains in amphibians (thought to have diverged from the main vertebrate evolutionary line about 200 million years ago) but not in sharks, in which all the species examined have had only one major chain class. These concordances show that our assumption of an approximately constant mutation rate in the immunoglobulin C regions is a reasonable one.

Homology of Immunoglobulins to Other Proteins

Surprisingly, there is significant homology between amino acid sequences of immunoglobulins and at least two other types of protein: the major histocompatibility antigens and the Thy-1 antigen, both discussed in Chapter 5. Both are membrane-bound glycoproteins, but only the histocompatibility antigens, which are important in defense against viral infections, play an established role in immune responses. These antigens consist of a 45,000 molecular weight heavy chain noncovalently bound to a smaller 12,000 molecular weight protein, β2-microglobulin. There is significant homology both in the position of disulfide bridges and in amino acid sequence (about 20% identity) between a typical immunoglobulin constant region domain, a β2-microglobulin sequence, and a region (residues 180–270) of the carboxy-terminal half of a typical histocompatibility antigen heavy chain. There is a similar degree of homology between constant region domains and the Thy-1 sequence, whose 111 residues also include a single immunoglobulin-like disulfide-bridged loop. Finally, the enzyme superoxide dismutase, although it shows no sequence homology to immunoglobulins, has a somewhat similar three-dimensional structure. Variable and constant regions are not present in the histocompatibility antigens, in Thy-1, or in this enzyme, as in the immunoglobulins, so the relevance of these homologies to the proteins' functions is unknown.

RECOMMENDED READINGS

Books

Eisen, H. N. (1974), *An Introduction to Molecular and Cellular Principles of the Immune Responses*. New York: Harper & Row.

Gergely, J. and Medgyesi, G. A. (1975), *Antibody Structure and Molecular Immunology*. Amsterdam: North Holland.

Glynn, L. E. and Steward, M. W. (1977), *Immunochemistry. An Advanced Textbook*. New York: Wiley.

Humphrey, J. H. and White, R. C. (1970), *Immunology for Students of Medicine* (3rd ed.). Oxford: Blackwell.

Kabat, E. A. (1976), *Structural Concepts in Immunology and Immunochemistry* (2nd ed.). New York: Holt, Rinehart & Winston.

Klein, J. (1982), *Immunology, The Science of Self–Nonself Discrimination* (1st ed.). New York: Wiley, pp. 156–255.

Landsteiner, K. (1945), *The Specificity of Serological Reactions* (2nd ed.). Cambridge, Mass.: Harvard University Press. (Reprinted in paperback, 1962, New York: Dover Publications).

Litman, G. W. and Good, R. A. (Eds.) (1978), *Comprehensive Immunology*. Vol. 5: *Immunoglobulins*. New York: Plenum Medical Books.

Manning, M. J. and Turner, R. J. (1976), *Comparative Immunobiology*. Glasgow: Blackie.

McConnell, I., Munro, A., and Waldman, H. (1981), *The Immune System* (2nd ed.). Oxford: Blackwell.

Nisonoff, A., Hopper, J. E., and Spring, S. B. (1975), *The Antibody Molecule*. New York: Academic Press.

Ogra, P. L. and Jacobs, D. M. (Eds.) (1982), *Regulation of the Immune Response*. Basel: S. Karger.

Reviews

Barker, W. C., Ketcham, R., and Dayhoff, M. O. (1978), Immunoglobulins. In *Atlas of Protein Sequence and Structure*, Vol. 5, Suppl. 3. Washington, D.C.: National Biomedical Research Foundation, p. 197.

Beale, D. and Feinstein, A. (1976), Structure and function of the constant regions of immunoglobulins. *Q. Rev. Biophys.* 9, 135.

Frangione, B. and Franklin, E. C. (1979), Split immunoglobulin genes and human heavy chain deletion mutants. *J. Immunol.* 122, 1177.

Koshland, M. E. (1975), Structure and function of the J chain. *Adv. Immunol.* 20, 41.

Pink, J. R. L. (1981), Immunoglobulins. In M. Gutfriend (Ed.), *Biochemical Evolution*. Cambridge: Cambridge University Press, p. 230.

Potter, M. (1978), Antigen-binding myeloma proteins of mice. *Adv. Immunol.* 25, 141.

Putnam, F. W. (1977), Immunoglobulins. In F. W. Putnam (Eds.), *The Plasma Proteins*, Vol. II. New York: Academic Press, p. 1.

Stanworth, D. R. and Turner, M. W. (1978), Immunochemical analysis of immunoglobulins and their subunits. In D. M. Weir (Ed.), *Handbook of Experimental Immunology* (3rd ed.). Oxford: Blackwell, p. 61.

Winkelhake, J. L. (1978), Immunoglobulin structure and effector functions. *Immunochemistry* 15, 695.

Articles

Campbell, D. G., Williams, A. F., Bayley, P. N., and Reid, K. B. M. (1979), Structural similarities between Thy-1 antigen from rat brain and immunoglobulin. *Nature* 282, 341.

Goding, J. W. and Shen, F-W. (1982), Structure of the murine plasma cell alloantigen PC-1: comparison with the receptor for transferrin. *J. Immunol.* 129, 2636.

Honjo, T., Obata, M., Yamawaki-Kataoka, Y., Kataoka, T., Kawakami, T., Takahashi, N., and Mano, Y. (1979), Cloning and complete nucleotide sequence of mouse immunoglobulin $\gamma 1$ chain gene. *Cell* 18, 559.

Kehry, M., Sibley, C., Fuhrman, J., Schilling, J., and Hood, L. E. (1979), Amino acid sequence of a mouse immunoglobulin μ chain. *Proc. Natl. Acad. Sci (USA)* 76, 2932.

Liu, Y. S. C., Low, T. L. K., Infante, A., and Putnam, F. W. (1976), Complete covalent structure of a human IgA1 immunoglobulin. *Science* 193, 1017.

Raison, R. L., Hull, C. J., and Hildemann, W. H. (1978), Characterization of immunoglobulin from the Pacific hagfish, a primitive vertebrate. *Proc. Natl. Acad. Sci. (USA)* 75, 5679.

Rogers, J., Early, P., Carter, C., Calame, K., Bond, M., Hood, L., and Wall, R. (1980), Two mRNAs with different 3′ ends encode membrane-bound and secreted forms of immunoglobulin μ chain. *Cell* 20, 303.

Roos, M. H. and Demant, P. (1982), Murine complement factor B (BF): sexual dimorphism and H-2-linked polymorphism. *Immunogenetics* 15, 23.

Sakano, H., Rogers, J. H., Hüppi, K., Brack, C., Traunecker, A., Maki, R., Wall, R., and Tonegawa, S. (1979), Domains and the hinge region of an immunoglobulin heavy chain are encoded in separate DNA segments. *Nature* 277, 627.

Torano, A., Tsuzukida, J., Liu, Y. V., and Putnam, F. W. (1977), Location and structural significance of the oligosaccharides in human IgA1 and IgA2 immunoglobulins. *Proc. Natl. Acad. Sci. (USA)* 74, 2301.

Trägårdh, L., Wiman, K., Rask, L., and Peterson, P. A. (1978), Amino acid sequence homology between HLA-A,B,C antigens, $\beta 2$-microglobulin and immunoglobulins. *Scand. J. Immunol.* 8, 563.

Tucker, P. W., Marcu, K. B., Newell, N., Richards, J., and Blattner, F. R. (1979), Sequence of the cloned gene for the constant region of murine $\gamma 2b$ immunoglobulin heavy chain. *Science* 206, 1303.

CHAPTER 3

The Genetics of Immunoglobulin Molecules

The introduction of a foreign antigenic material into any vertebrate organism rapidly results in the development of an immune response specifically directed toward that particular antigen. This remarkable specificity of the immune response to each different antigen presents one of the most intriguing problems in current immunology.

The immune response can be divided into cellular and humoral immunity, the latter being characterized by a lymphocytic infiltration and the production of specific antibody molecules. (The cellular aspect of the immune response is discussed in Chapter 5.) This chapter discusses the genetic control of the synthesis of the antibody molecules.

The enormous heterogeneity of the immunoglobulin molecules has long puzzled geneticists. Two major events greatly increased our still incomplete understanding of this problem. The first was the demonstration of genetic polymorphism of antigenic determinants (i.e., "allotypic" markers) on the immunoglobulin molecules themselves. These allotypic markers have provided a new approach to problems in formal, population, biochemical, and medical genetics.

The second major event was the establishment of the "two genes for one polypeptide chain" hypothesis for immunoglobulin chains. In contrast to the conventional "one gene—one polypeptide chain" mechanism governing the synthesis of almost all other proteins examined, each immunoglobulin polypeptide chain is the product of at least two different structural genes, one for the variable (V) region and another for the constant (C) region of the chain. Furthermore, in heavy (H) chains, different V-regions subgroups can be found in association with different C-region classes (and subclasses). This reciprocal sharing of V- and C-region genes plus the random association of light (L)-chain types with H-chain classes (and subclasses) in immunoglobulin molecules is one possible mechanism of generating heterogeneity for immunoglobulins from a relatively limited number of genes.

During the normal course of an immune response in mammalian species, the appearance of IgM molecules precedes the appearance of IgG molecules. Apparently, identical H-chain V regions may be associated with different H-chain C regions. It appears likely that a single V region may be expressed initially on μ chains and later in the immune response on other classes of H chains (e.g., γ, α, ϵ).

Perhaps the most interesting aspects of immunogenetics are allelic exclusion, specific gene activation, and differential gene expression. During maturation of immunocytes one allele of a given gene is specifically activated among a large number of similar genes for immunoglobulin polypeptide chains. This observation demonstrates that allelic exclusion is not limited to sex chromosomes; instead it may well be one of the basic mechanisms controlling differentiation. Furthermore, not all functionally related genes are expressed in a given cell at a given time or perhaps even in a given species.

THE "MULTIPLE GENES—ONE POLYPEPTIDE CHAIN" CONCEPT

History

In 1941 Beadle and Tatum proposed the "one gene–one enzyme" hypothesis. The synthesis of every protein studied, in microorganisms as well as in higher organisms, was thought to follow the general rule that one gene controls the synthesis of a single polypeptide chain. However, in 1965 Dreyer and Bennett suggested that two genes might be involved in the synthesis of a single immunoglobulin chain. Their hypothesis was based on results of peptide mapping of Bence–Jones proteins and myeloma protein L chains derived from inbred mice. Simultaneously two groups, one led by Hilschmann and the other by Putnam, independently showed by amino acid sequences of three human Bence–Jones proteins that the carboxy-terminal half of these chains was always identical within a given type of L chain (with the exception of allotypic and isotypic differences) but that each of these L chains contained a unique set of peptides in the amino-terminal half that distinguished each L

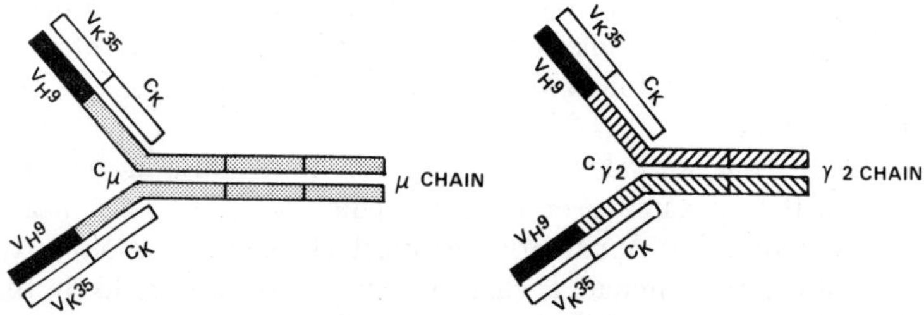

Fig. 3.1 Schematic representation of the structural data for a monoclonal IgM(κ) and a monoclonal IgG2(κ) isolated from a single patient (Til). The two molecules have identical L chains and identical V_H regions but their C_H regions have less than 30% amino acid sequence homology.

chain from any of the others. These findings led to the suggestion that two structural genes, one for the carboxyl half of the chain and the other for the amino half, controlled the synthesis of each immunoglobulin L chain. Fusion of the C- and V-region genes by some as yet unknown mechanism (probably a deletion) is thought to complete the genetic information necessary for the synthesis of each chain.

The concept that multiple genes encode a single immunoglobulin polypeptide chain is consistent with the observation that, in H chains, different V-region subgroups can associate randomly with C regions of all classes and subclasses. This concept also offers a logical explanation for Todd's observation that, in rabbits, various so-called group "a" allotypes (e.g., Aa1, Aa2, Aa3),* which are markers of V_H regions, were present on IgG, IgA, and IgM.

At the protein level, the best evidence for the hypothesis that the V and C regions of each immunoglobulin polypeptide chain are encoded by separate genes comes from amino acid sequences of Ig polypeptide chains, and this can be exemplified by the work on serum from a patient (Til) with a double paraproteinemia. This serum contained two monoclonal proteins, one an IgG2(κ) and the other IgM(κ). The two molecules have identical L chains, and the V regions of their H chains are also identical, but the C region of the γ2 chain differs from that of the μ chain by over 70% of their amino acid sequences (Fig. 3.1). The C region of the γ2 is considerably smaller than that of the μ chain (325 versus 450 amino acid

* These allotypes are not allotypes of a single structural gene; they may actually represent sets of closely linked isotypes (see p. 87).

residues), but the V region of both chains contains 115 amino acid residues. Genetic analysis indicated that the C regions of $\gamma 2$ and μ chains are encoded by distinct genes. Evolutionary studies suggest that $C\gamma$ genes diverged from the $C\mu$ gene approximately 200 million years ago. There is no known genetic mechanism that would allow two genes to maintain an identical stretch of 345 nucleotides through a period of 200 million years in evolution and simultaneously permit the remaining 975 or more nucleotides of the same two genes to diversify to the extent reflected by 70% difference in the polypeptide chain coded by them. The only logical explanation is to assume that the V and the C regions of each H chain are encoded by separate genes.

Arrangement and Rearrangement of Immunoglobulin L-Chain Genes

Direct proof for the "multiple genes for one polypeptide chain" hypothesis finally came from DNA sequencing, gene cloning, and DNA/RNA hybridization studies in the late 1970s. Structural and genetic analyses demonstrated that the basic unit of each Ig molecule is encoded by at least four genes: V_H, C_H, V_L, and C_L. (Nucleotides encoding the J segment of both H and L chains and the D segment of the H chain are also considered as minigenes by some immunologists.) A normal individual has large numbers (hundreds or more) of V_H and V_L genes, and moderate numbers of C_H and C_L genes. These genes belong to three linkage groups, as illustrated in Fig. 3.2. In mice, the H-, K-, and λ-chain genes are located on chromosomes 12, 6, and 16 respectively. In humans, the H-chain genes are on chromosome 14, whereas the κ- and λ-chain genes are on chromosomes 2 and 22, respectively.

Nucleic acid analyses have shown that within each linkage group the V genes are far apart from the C genes on the chromosome in germ and embryonic cells. However, in differentiated antibody-producing cells, one of the V genes is translocated to the neighborhood of a C gene, for the synthesis of a single polypeptide chain.

Further nucleic acid studies have shown that Ig genes are split genes consisting of *exons,* which are transcribed and eventually translated into amino acid sequences, and *introns,* which are transcribed during synthesis of a messenger RNA precursor but not translated into amino acid sequences. Each domain in an Ig polypeptide chain is encoded by an exon. The hinge region of a γ chain and the tail

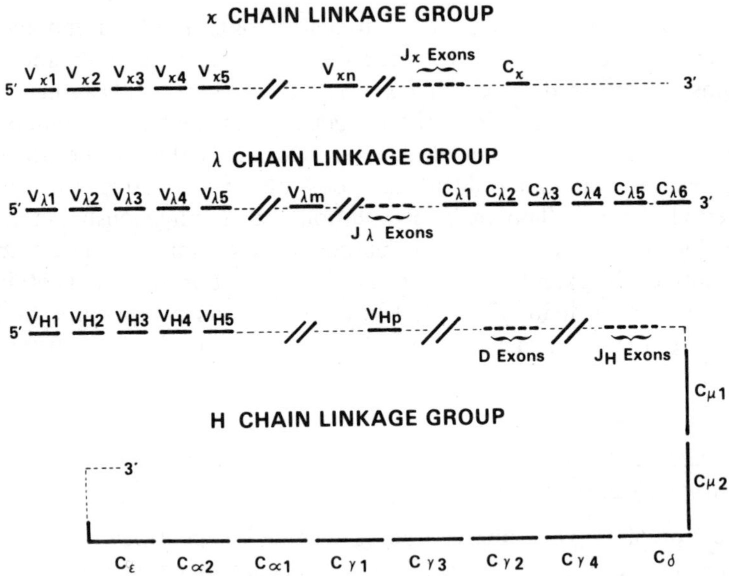

Fig. 3.2 Schematic representation of human Ig structural genes (based on more extensive work on mouse Ig genes). Each J exon encodes approximately 11 to 13 amino acid residues at the carboxy-terminal end of the V region. Each D exon encodes a short stretch of amino acids equivalent to the third hypervariable region of the H chain. The order of the C-region genes is arbitrary.

of membrane μ and δ chains are also encoded by individual exons. In the DNA, the borders of the introns are usually marked by GT and AG sequences. The genetic events leading to the synthesis of an Ig chain are outlined in Fig. 3.3. In germ and embryonic cells the nucleotides encoding a full L chain are located in two large DNA fragments, which are far apart. One of the fragments contains the leader (L) and the V exons, which are separated by an intron. The second DNA fragment contains the J and C exons, which are separated by another intron. During differentiation of Ig-producing cells, a somatic translocation joins the V to the J. This translocation commits the cells and their progeny to the synthesis of the L chain specified by the particular V, J, and C exons involved. In differentiated cells, all deoxynucleotides from the leader exon to the C exon are transcribed into the precursor messenger RNA (mRNA), which is processed inside the nucleus to delete the introns before becoming a mature mRNA. The mRNA then goes to the polyribosome to be translated into a L-chain precursor. A postsynthetic cleavage cuts off

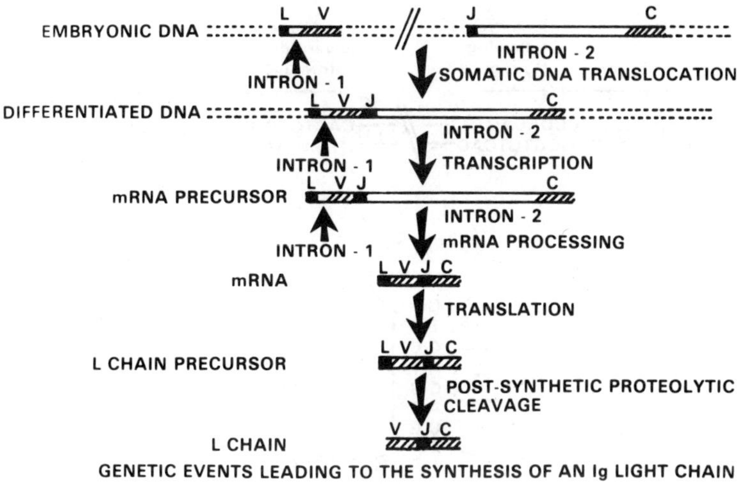

Fig. 3.3 Genetic events involved in the production of a light chain. Shaded areas represent exons; white bars represent introns. The L exon codes for the leader piece.

the leader piece from the NH_2-terminal end of the precursor L chain and turns the precursor L chain into a mature L chain.

Theoretically, C and V regions may fuse at one of three levels: DNA, mRNA, or protein. Protein level association has been excluded, for it has been demonstrated that each H or L chain is synthesized at one point as one unit from its amino-terminal end on the polyribosomes. Furthermore, nucleotide sequence studies by Milstein on a purified mouse L-chain mRNA indicate that a single mRNA molecule codes for both the V region and the C region of the L chain.

Fusion at the level of mRNA has not been totally excluded, but a majority of the experimental evidence is consistent with a "loop-out" deletion model at the DNA level. Figure 3.4 illustrates such a model using a mouse κ-chain system as an example. DNA-sequence studies revealed certain unusual stretches of DNA with palindrome (inverted repeat) sequences in the introns after the V exon and before the J exon. The complementary relation between the palindrome sequence allows them to become aligned to bring together widely separated V and J exons. It is believed that the paired palindromes also serve as recognition sites for an enzyme (e.g., recombinase) that catalyzes the cleavage and splicing steps, and that during these steps the V–J segments are fused and the nucleotides in between are deleted.

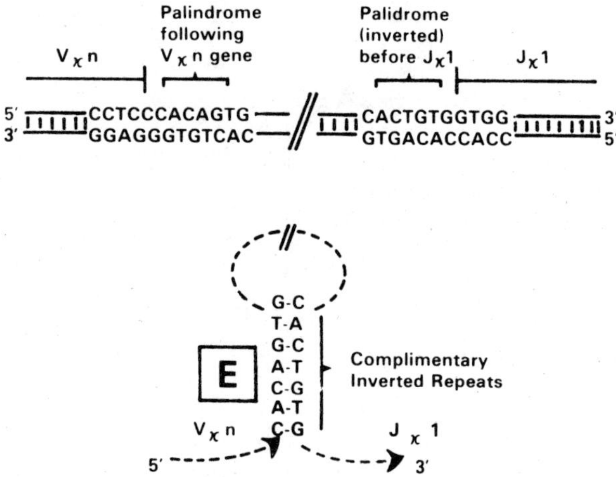

Fig. 3.4 A possible mechanism for the joining of a $V\kappa$ gene to a $J\kappa$ gene. The complementary inverted repeats (palindromes) bring together widely separated V and J segments. A hypothetical enzyme (e.g., recombinase) then catalyzes the splicing event.

The Genetic Switch

During the course of a primary immunization, the antibodies produced at the early stage usually belong to the IgM class. Subsequently, the IgM antibodies are gradually replaced by IgG antibodies. The early IgM and the late IgG antibodies against a single antigen often have common idiotypic specificities. The genetic mechanism for the IgM to IgG switch was not known until the early 1970s. In our studies of serum from a patient (Til) with double paraproteinemia, we found two monoclonal immunoglobulins: an IgG2(κ) and a IgM(κ). The two molecules have identical V_H, V_L, and C_L regions, but different C_H regions (Fig. 3.1). Immunological studies showed that the Til IgM and the Til IgG were secreted by two subclones of closely related antibody-producing cells (Fig. 3.5) Rhondamine-labeled anti-μ stained IgM-producing cells only, and fluorescein-labeled anti-γ stained IgG-producing cells only, but fluorescein-labeled antiidiotype antibodies reacted with both populations of cells.

On the basis of combined biochemical and immunological data, we proposed a "genetic switch" hypothesis. This hypothesis suggests that during B-cell maturation, certain IgM-producing cells undergo

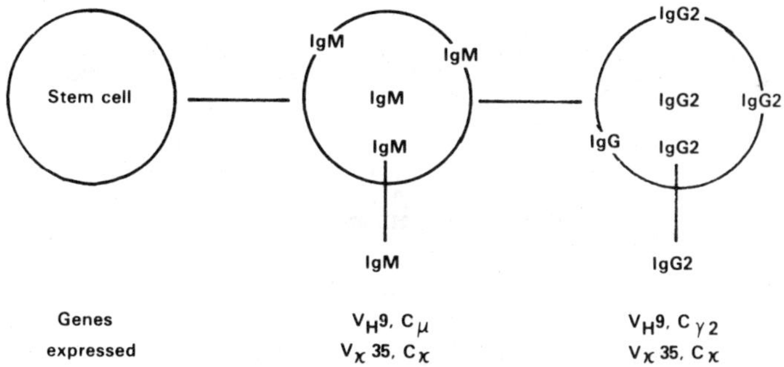

Fig. 3.5 A diagrammatic representation of cytodifferentiation of immunocytes showing the switch from an IgM-producing cell to an IgG-producing cell.

a switch to the production of Ig belonging to another class (Fig. 3.5). During the switch a given V_H gene, which was associated with a $C\mu$ gene for the production of IgM, is translocated to another C_H gene for the synthesis of a new Ig class. Antibodies of the new class retain the same idiotype and antigen-binding specificity as the parental IgM. This hypothesis was subsequently verified by nucleic acid analyses. In 1980, seven laboratories independently reported that two types of DNA rearrangements are involved in the synthesis of H chains other than the μ chain. The first rearrangement is equivalent to the V–J translocation of the L chain but may require two steps (in which a V_H exon is joined to one or more D exons first, then the V_H–D piece is joined to a J_H exon adjacent to the $C\mu$ gene). The second rearrangement, the C_H switch, replaces the $C\mu$ gene with another C_H gene. A schematic outline of genetic events involved in H-chain production is shown in Figure 3.6. The deletion of the $C\mu$ gene (and other C_H genes) might accompany the C_H switch. It is generally believed that the initial V–J translocation leads to IgM production, whereas the C_H switch leads to the production of IgG, IgA, or IgE. The possible mechanisms leading to IgD synthesis are not yet fully understood; they may involve either a C_H switch or a μ-δ mRNA processing, or both. A diagrammatic illustration of the two genetic mechanisms for IgD gene expression is presented in Figure 3.7.

Recent evidence suggests that DNA sequences termed enhancers are responsible for expediting the transcription of genes into mRNA. These enhancers all seem to possess at least the common sequence GTGGXXXG, where G is guanine, T is thymine, and X is either adenine or thymine. These enhancers may help solve the enigma of specific Ig gene activation in that they exist in the intron between

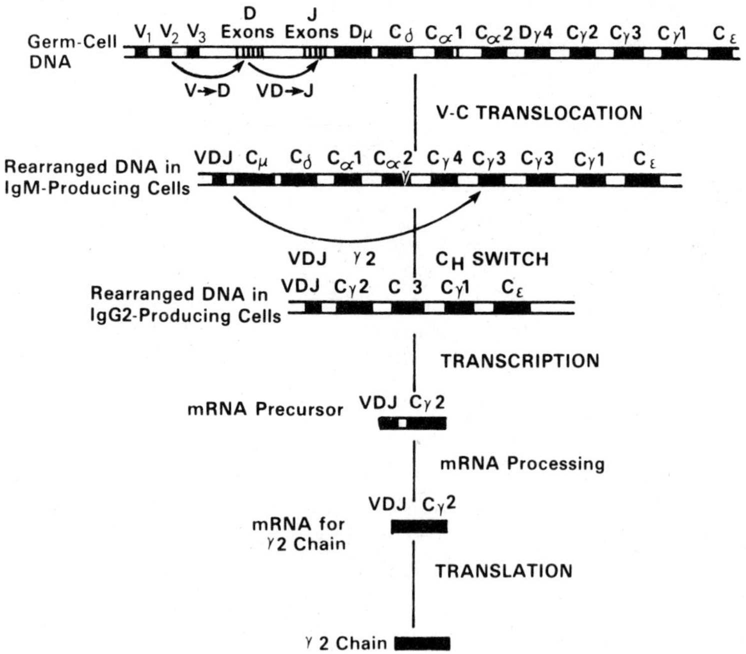

Fig. 3.6 A schematic illustration of the genetic events leading to the production of a γ2 chain in humans. Two rearrangements of Ig genes are apparently involved: a V–C translocation and a CH switch. In humans, the H-chain genes are located on chromosome 14, and there is evidence that a Cα gene is probably at the 5′–OH side of the Cγ gene. The situation differs in the mouse system, in which the H-chain genes are located on chromosome 12, and the Cα gene is on the 3′–OH side of the Cγ gene.

the V-D-J coding region and the C region exons of the rearranged heavy chain gene. For a DNA segment to be characterized as an enhancer, it must be:

1. essential for transcription
2. moved around or inverted
3. work in heterologous genomes.

Although the exact mechanism by which enhancers stimulate transcription is unknown, it has been speculated that inappropriate enhancer expression could be a potential factor in the neoplastic transformation process (Marx, 1983).

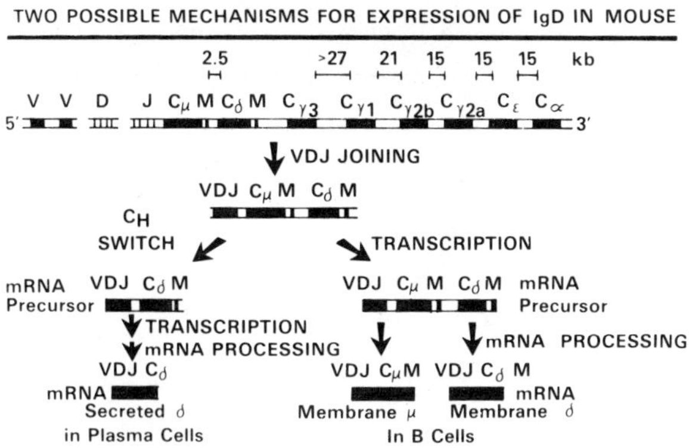

Fig. 3.7 Two possible mechanisms for expression of IgD in mice.

ALLOTYPIC MARKERS OF HUMAN IMMUNOGLOBULINS

Most early information about the genetic control of immunoglobulins was obtained from studies of so-called allotypic markers, which are inherited antigenic determinants on the immunoglobulin molecules, detected by immunological methods. Such antigenic determinants have been identified in most mammals (e.g., human beings, monkeys, rabbits, cattle, mice, and rats) and several other species (e.g., chickens). It is likely that they occur in all vertebrates that have an immune system and are capable of mounting an antibody response against "foreign" antigens.

Nomenclature

Five sets of allotypic markers are recognized on human immunoglobulins, the genetically closely linked Gm, Am, and Mm systems, and the independently inherited Hv and Km (or Inv) systems. These allotypic markers have also been termed "factors" (i.e., Gm factor, Inv factor). Gm, Am, and Mm indicate genetic markers on heavy chains of IgG, IgA, and IgM, respectively, Hv indicates an inherited antigenic determinant on the V region of H chains, and Km indicates a genetic marker on κ-type L chains (Table 3.1). The term *Km* was recommended by an International Committee on Immunoglobulin Allotypes in 1974 to replace the old terminology *Inv*. The same committee also adopted the terminology initially used by

Table 3.1 Nomenclature of allotypic markers of human immunoglobulins

WHO	Original	WHO	Original
G1m(1)	a	Gm(7)	γ
G1m(2)	x	Gm(8)	e
*G1m(3) or (4)	*f, b^w or b^2	Gm(9)	P
G1m(17)	z	Gm(12)	$b\gamma$
		Gm(18)	Ro^2
G2m(23)	n	Gm(19)	Ro^3
		Gm(20)	z (S.F.)
G3m(5)	*b, b^1		
G3m(6)	*c^3, like	A2m(1)	A2m(1)
G3m(10)	b^5	A2m(2)	A2m(2)
G3m(11)	b^0		
G3m(13)	b^3	Mm(1)	Mm(1)
G3m(14)	b^4		
G3m(15)	s	Km(1)	Inv(1)
G3m(16)	t	Km(2)	Inv(2)
G3m(21)	g	Km(3)	Inv(3)
G3m(24)	c^5		
G3m(26)	u	Hv(1)	Hv(1)
G3m(27)	v		
G3m(28)	G3m(28)		

* Refer to the same antigenic determinant(s).

van Loghem, Wang, and Shuster, in which a number was inserted immediately before the small letter *m* to indicate the subclass specificity. For example A2m, G1m, G2m, and G3m indicate that these markers are associated with IgA2, IgG1, IgG2, and IgG3 subclasses, respectively. Table 3.1 lists the currently known factors in each system. The alphabetical nomenclature was originally used by researchers working in this field, whereas the numerical nomenclature was recommended by a committee of the World Health Organization to reduce the confusion that inevitably arises when a number of laboratories are investigating an inherently complex phenomenon. Both notations have been widely used, and either is acceptable at this time.

The allotypic determinants are expressed as letters or numbers in parentheses, for example, G1m(a) or G1m(1). The alleles are letter or number superscripts at the right upper corner of the corresponding Gm, Am, or Km system, that is, Gm^a or Gm^1. Some factors originally thought to be distinct were subsequently shown not to be so; for example G1m(3) is identical to G1m(4).

History

In 1939 Waller reported that the sera of certain patients with rheumatoid arthritis agglutinated sheep red blood cells coated by low (subagglutinating) doses of rabbit anti-sheep red blood cell antibody. The factor causing this agglutination was designated "rheumatoid factor." Subsequently sera containing rheumatoid factor were shown to have antibody toward both human and rabbit IgG. Rheumatoid factor in sera of patients with rheumatoid arthritis may be heterospecific or isospecific. Heterospecific rheumatoid factors react with all human IgG; isospecific rheumatoid factors react with IgG from human sera of a given allotype other than that of the individual from whom the "anti-isoglobulin" is obtained.

In 1956, during investigations on rheumatoid factor, Grubb noted that sera from some patients with rheumatoid arthritis agglutinated Rh-positive red blood cells coated with certain "incomplete" (nonagglutinating) anti-Rh sera but did not agglutinate cells coated with other anti-Rh sera of equivalent or greater titer. The agglutinating system was inhibited by some human sera but not by others. The inhibitory activity present in normal sera resided in the immunoglobulins, and the factor responsible for the inhibition was designated Gm(a) [which was subsequently changed to G1m(1) according to the currently recommended nomenclature]. The G1m(1) inhibitory activity was shown by Grubb and Laurell to be inherited as a simple Mendelian genetic autosomal trait.

It was later shown that antibodies with anti-immunoglobulin activity are not restricted to patients with rheumatoid arthritis. Such antibodies (antiglobulins) exist in sera of patients with certain other diseases, as well as in sera of some normal individuals. In normal individuals the antibodies presumably arise as a result of sensitization, by transfusion or pregnancy, to human immunoglobulins containing genetically "foreign" allotypic determinants. These nonrheumatoid antibodies are now used in most standard tests for the presence of allotypic determinants, because they give better differentiation between serum positive and serum negative for Gm factors than rheumatoid anti-Gm sera do.

Techniques for Testing

Allotypic markers in animals other than human beings are detected either by direct precipitation reactions or by hemagglutination inhi-

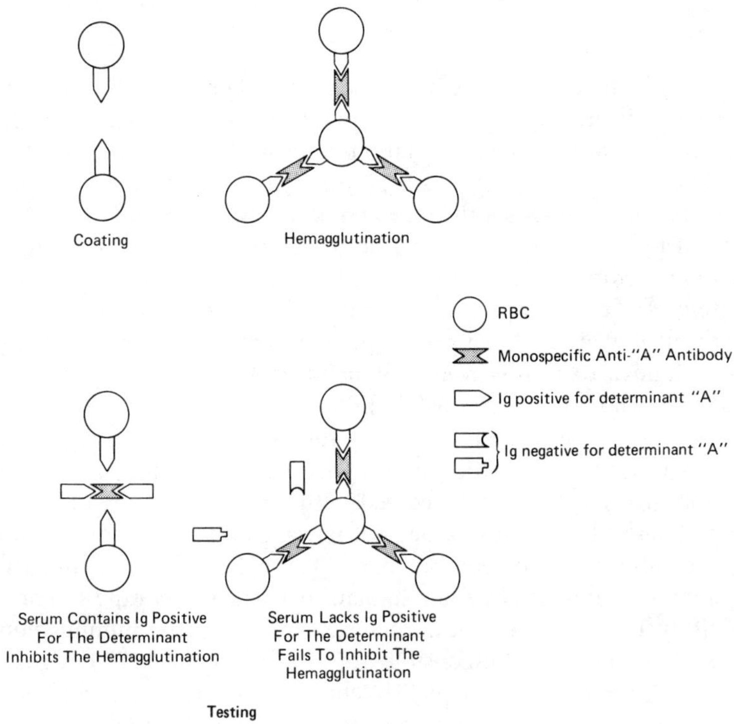

Fig. 3.8 Diagrammatic illustration of the hemagglutination inhibition method for the detection of antigenic determinants.

bition reactions similar to those described by Grubb. The former method is simple and convenient; unfortunately, because of the rarity of precipitating antisera to human allotypes, precipitation reactions can be used only occasionally to detect a few of the allotypes [i.e., G1m(17), G2m(23), and G3m(21)] in human systems. Most human allotypes are thus routinely typed by the hemagglutination procedure. An outline of this procedure is given in Figure 3.8.

A test system for detecting immunoglobulin allotypic determinants contains four main components:

1. Rh-positive red blood cells
2. An immunoglobulin known to be positive for the allotypic marker to be tested, to be coated onto the red blood cells either by immunological methods (incomplete anti-Rh anti-

bodies) or nonimmunological methods (e.g., tannic acid or chromic chloride)

3. The specific antibody (antiglobulin) directed against the allotypic marker to be tested
4. The test sample, usually serum or isolated immunoglobulin or immunoglobulin fragments

In standard test systems for each allotypic marker, Rh-positive red blood cells coated with immunoglobulin of a given Gm, Am, or Km specificity are used as antigen, and human antiglobulins specific for that factor are used as antibody. Mixing of the two results in agglutination of the cells. An immunoglobulin (whether in the isolated form, as an isolated chain or fragment, or in whole serum) positive for the factor inhibits the agglutination by reacting with the anti-allotype antibodies and making them unavailable to react with the immunoglobulins bound to the red cells, whereas proteins lacking the factor fail to inhibit.

Alloantisera

As stated above, antiglobulins from normal sera (called "Snaggs," for serum normal agglutinators) are now generally used as typing reagents for detection of human allotypic markers. In contrast to "rheumatoid factor" agglutinators (called "Raggs," for rheumatoid agglutinators), Snaggs are usually monospecific for only one allotypic marker and provide better discrimination between inhibitory and noninhibitory normal human sera.

Anti-Gm agglutinators are frequently present as a transient phenomenon in normal infants. Indeed experimental evidence strongly suggests that all infants who lack a Gm factor present in the mother produce anti-Gm antibodies specifically directed toward one or another maternal Gm factor. The antibodies first appear during the latter part of the first year of life and gradually disappear by the age of 3 or 4 years. However, about 1 to 2% of normal individuals without known disease or exposure to transfusion have detectable anti-Gm antibodies in their serum, an incidence suggesting that in some individuals these antibodies persist throughout life. Because the fetus is exposed to maternal gammaglobulin during intrauterine life, the invariable formation of anti-Gm antibodies shortly after birth is of great relevance in evaluating data on immune tolerance in human beings. These findings suggest that allotypic determinants cannot induce tolerance in a situation of human maternal–fetal incompatibility.

Since IgG of the fetus appears to cross the placenta into the maternal circulation, anti-Gm agglutinators can arise by immunization of the mother by fetal immunoglobulin. This occurs in the fetus that has inherited from the father Gm alleles not present in the mother's genome; for example, if the fetus of a G1m(1—) mother inherits a G1m(1) allele from the father, G1m(1) globulin may enter the maternal circulation, and the mother may be immunized by the G1m(1) antigen. This mechanism for anti-Gm formation is thus analogous to formation of anti-Rh antibodies in Rh-negative mothers of Rh-positive infants.

In addition to human anti-Gm antisera, reagents produced in other species have been used by several groups. Rhesus monkeys (which lack most known Gm factors), when immunized with pooled human IgG, occasionally produce anti-human immunoglobulin reagents that are useful in Gm typing. Similarly rabbit antisera produced to myeloma proteins positive for only one Gm factor and subsequently absorbed with myeloma proteins lacking this factor often provide valuable anti-Gm reagents. Indeed such reagents led to the discovery of factors G1m(17), G2m(23), and G3m(21).

Whereas most antibodies are of the IgG class, human antibodies directed toward human immunoglobulins are almost always of the IgM class. Why this is so is puzzling. The nature of the antigenic determinant and the route of immunization may play a role. Moreover, the species from which the antibodies are produced may be a factor. (As discussed below, rabbit anti-rabbit allotype antibodies are usually of the IgG class.)

Location of Allotypic Determinants on Immunoglobulin Molecules

L-Chain determinants. The Km factors are associated with κ-type L chains of immunoglobulins, and, since L chains are common to all classes of immunoglobulins, Km activity can be found in association with IgG, IgA, IgM, IgD, and IgE molecules. Extensive studies of myeloma proteins with known L-chain types have demonstrated that the Km factors are carried by κ-type L chains but not by λ-type L chains.

The activities of Km factors depend in part on the quaternary structure of the immunoglobulin molecule for their expression. Isolated L chains require the presence of five times the molar concentration of intact immunoglobulins for the detection of Km activities. In studies on myeloma proteins either the Km(1) factor was undetectable in IgG2 and IgG4 molecules, or much higher protein concentrations were required for its detection in such molecules. Recombinants of isolated Km(1)-positive L chains with $\gamma2$ chains showed less Km(1) activity than did the original protein or recombinant molecules formed by the same Km(1)-positive L chains and $\gamma1$ or $\gamma3$ chains. This observation indicates that $\gamma2$ and $\gamma4$ chains, when combined with L chains, give rise to a quaternary structure in which the Km(1) determinant is poorly expressed.

H-Chain determinants. Gm factors are associated with IgG molecules only. All Gm factors found to date have been located on the C regions of γ chains, and the activity of many has been localized to the Fc fragments of these chains. However, two factors, G1m(4) and G1m(17), were found to be located on the Fab fragment (more precisely, the C region of the Fd fragment). These two factors are not usually detectable on either isolated L or H chains but require the presence of both in a reconstituted form. The conclusion that G1m(4) and G1m(17) determinants are located on H chains is based on the observation that H chains could restore the corresponding Gm specificity regardless of which type of L chain they were recombined with. Therefore, G1m(4) and G1m(17) can be regarded as hybrid determinants that actually appear as serological entities only when H and L chains are combined.

The Gm specificities in IgG of pooled normal human serum are not present on all the IgG molecules in the serum. An early clue was the finding that G3m(5) activity per unit weight of isolated IgG3 myeloma proteins was several times higher than the pooled IgG of whole normal human serum. Gm typing on large numbers of human myeloma proteins demonstrated a correlation between certain Gm markers and IgG subclasses. As shown by the WHO-recommended numerical nomenclature in Table 3.1, G1m(1), G1m(2), G1m(4), and G1m(17) associate with IgG1 subclasses only; G2m(23) with IgG2 subclasses only; and G3m(5), G3m(21), and several others with IgG3 subclasses only.

These allotypes reflect genetic heterogeneity of immunoglobulin molecules; thus not all allotypic markers found in a given subclass are present in all immunoglobulins of that subclass. To illustrate, all IgG3 myeloma proteins carry either G3m(5) or G3m(21) factor, but not both, whereas all IgG1 myeloma proteins from Caucasians and Negroes can be positive for G1m(1) or G1m(4), but not for both.

Chemical Basis of Allotypic Specificities

Km factors. Immunoglobulins are glycoproteins; theoretically allotypic markers of immunoglobulins could be ascribed to the primary structure of the polypeptide part of the molecule, to its carbohydrate moiety, or to a combined peptide–carbohydrate entity. Peptide mapping and amino acid sequence analyses have shown that Km factors are located on the C region of the L chains. As shown in Table 3.2, κ chains positive for Km(1) factor have valine at position 153 and leucine at position 191; those positive for Km(3) have alanine at position 153 and valine at position 191; and those positive for both Km(1) and Km(2) have alanine at position 153 and leucine at position 191. The remaining portions of the C regions of these κ chains appear to have an identical amino acid sequence. These studies clearly demonstrate that the antigenic determinants for various Km factors differ from one another by a single amino acid residue. Examination of genetic code indicates that both the alanine–valine interchange and the leucine–valine interchange can be the result of single-step point mutation.

G1m(1) factor. The same seems to be true for most Gm factors. One example comes from studies done on the relationship of amino acid sequence to the G1m(1) factor. Peptide maps showed that in human IgG1 H chains, a given peptide was always present when the G1m(1) factor was present, whereas another peptide was always present when the protein was G1m(1—). These two peptides were subsequently designated the "*a*" and "non-*a*" peptides.

Amino acid sequence studies showed that the non-*a* peptide is

Table 3.2 Amino acid substitutions responsible for various Km antigenic determinants

Km	Residue 153	Residue 191
1	Val	Leu
1,2	Ala	Leu
3	Ala	Val

located at the C_H3 homology region (residues 356 to 360 of the γ chain*). The amino acid sequence of the *a* peptide is established as Asp-Glu-Leu-Thr-Lys, and the sequence of the non-*a* peptide, as Glu-Glu-Met-Thr-Lys. Two amino acid substitutions were observed; hence the evolution of this Gm allotype is not a straightforward example of divergence by incorporation of a single point mutation.

Evolutionary study of G1m(1) revealed that this factor is present in some of the apes but absent in any species lower than the Old World monkeys. All apes that are G1m(2) positive have the *a* peptide, indicating that the correlation between the G1m(1) factor and the *a* peptide also holds in nonhuman primates. The Old World monkeys, which are all G1m(1—), have neither the *a* nor the non-*a* peptide but have another unique peptide designated the "OWM peptide." The amino acid sequence of the OWM peptide is established as Glu-Glu-Leu-Thr-Lys.

It appears that both the *a* and the non-*a* peptides were derived from the OWM peptide by single-step point mutations. One mutation that converted a glutamic acid residue in the OWM peptide into an aspartic acid residue gives rise to the *a* peptide, and another mutation that converted a leucine residue on the OWM peptide into a methionine residue gives rise to the non-*a* peptide. Examination of the genetic code indicated that both mutations could have resulted from single base changes:

Leu (CUG)	OWM peptide	(GAA)
↓	↙ ↘	Glu (GAG)
Met (AUG)	(non-*a* peptide) (*a* peptide)	↓ (GAU)
		Asp (GAC)

These findings suggest that the mutations that give rise to the *a* and non-*a* peptides from the OWM peptide were recent evolutionary events, occurring less than 20 million years ago.

Other factors. Other amino acid interchanges probably responsible for allotypic differences in immunoglobulin H chains in human beings as well as in some other species are summarized in Table 3.3. Arginine was found at position 214 of G1m(4) $\gamma 1$ chains, and lysine was found at the same position of G1m(4—) $\gamma 1$ chains. According to the genetic code, only one nucleotide replacement is required

* All γ-chain numbering is based on the sequence of the protein Eu studied by Edelman and colleagues.

Table 3.3 Amino acid interchanges probably associated with allotypic differences in immunoglobulins

Chain	Residue number*	Amino acid residue	Associated allotype
Human $\gamma 1$	356, 358	Asp, Leu	G1m(1)
		Glu, Met	G1m(1−)
Human $\gamma 1$	431	Gly	G1m(2)
		Ala	G1m(2−)
Human $\gamma 1$	214	Arg	G1m(3)
		Lys	G1m(3−)
Human $\gamma 3$	296, 436	Phe, Phe	G3m(5)
		Tyr, Tyr	G3m(21)
Human $\alpha 2$	212, 221	Ser, Arg	A2m(2)
		Pro, Pro	A2m(1)
Human κ	153, 191	Val, Leu	Km(1)
	153, 191	Ala, Leu	Km(1,2)
	153, 191	Ala, Val	Km(3)
Human H	V Region	Multiple differences	Hv(1), Hv(1−)
Mouse α	CH1 Domain	Bond between	In NZB mice
		H and L chains	In BALB/c mice
		No bond	
Mouse $\gamma 2a$	135	Ser	In C3H mice
		Thr	In BALB/c mice
Mouse κ	C Region	Multiple differences	b4, b5, b6, b9
Rat κ	C Region	Multiple differences	RL-1a, RL-1b
Rabbit γ	225	Met	d11
		Thr	d12
Rabbit γ	309	Thr	c14
		Ala	c15
Rabbit H	V Region	Multiple differences	a1, a2, a3

* Residue numbering is by homology with a human $\gamma 1$ chain, EU.

for the Arg ⇄ Lys interchange. Similarly, G1m(2) $\gamma 1$ chains differ from G1m(2—) $\gamma 1$ chains by a glycine to alanine substitution at position 431. According to the genetic code, the glycine to alanine interchange can also be explained on a single-step point mutation basis.

Similar studies done on IgG3 immunoglobulins have shown that two phenylalanine residues are found at positions 296 and 423, respectively, on G3m(5) $\gamma 3$ chains, whereas two tyrosine residues are found at these two positions on G3m(21) $\gamma 3$ chains. Again the Phe ⇄ Tyr substitution requires only one base change in the triplet of a codon.

Table 3.3 shows that immunoglobulins have two kinds of allotypes, called simple and complex. Simple allotypes, e.g., G1m(1), are associated with only one or a few amino acid substitutions. In contrast, complex allotypes, e.g., Hv(1), are associated with multiple (or often over 10%) amino acid differences.

Isoallotypes. Most allotypic antigenic determinants are restricted to a single subclass of H chain or a single subtype of L chain (Table 3.1). There are, however, some antigenic determinants that behave as allotypes within a specific subclass (or subtype) but also appear in all members of at least one other subclass (or subtype). These determinants are called "isoallotypic" determinants (formerly "non"-markers). The human isoallotypic determinants are summarized in Table 3.4. For example, isoallotypic determinant nG1m(1) acts as an allele of G1m(1) on IgG1 proteins, but the non-(1) determinant is also present on all IgG2 and IgG3 proteins. Similarly isoallotypic determinant nG4m(a) acts as an allele of nG4m(b) on IgG4 proteins, but G4m(a) is always found on IgG1 and IgG3 proteins, whereas G4m(b) is always found on IgG2 proteins. At least one isoallotype designed nA2m(2) was described from human IgA proteins. The nA2m(2) antigenic determinant occurs on all IgA1 mole-

Table 3.4 Distribution of isoallotypic determinants in immunoglobulin subclasses

Alphameric	Numeric	IgG1	IgG2	IgG3	IgG4	IgA1	IgA2
nG1m(a)	nG1m(1)	+1/3 −a/1	+	+	−*	−	−
nG1m(z)	nG1m(17)	+1/3 −z/17	−	+	+	−	−
nG3m(b0)	nG3m(11)	+	+	+g/21 −b0/11	−	−	−
nG3m(b1)	nG3m(5)	+	+	+g/21 −b1/5	−*	−	−
nG3m(g)	nG3m(21)	−	+	−g/21 +b1/5	−	−	−
nG4m(a)	nG4m(a)	+	−	+	+a −b	−	−
nG4m(5)	nG4m(b)	−	+	−	−a +b	−	−
nA2m(2)	nA2m(2)	−	−	−	−	+	+A2m(1) −A2m(2)

* Not detectable with all nG1m(a) specific reagents.

cules but behaves as an allotype among IgA2 molecules. The frequent occurrence of isoallotypes is consistent with the idea that gene expansion by means of gene duplication is an important mechanism governing the evolution of immunoglobulin structural genes.

Molecular Analysis of Human IgA

Human IgA molecules are classified into two subclasses, IgA1 and IgA2. The later may have either A2m(1) or A2m(2) allotype. In 1979, Putman and his associates determined the complete amino acid sequence of three human α chains: an α1 chain, an α2 chain of the A2m(1) allotype, and an α2 chain of the A2m(2) allotype. Comparison of these sequences provides an elegant example of the biochemical basis of antigenic determinants of the α-chain C region. Excluding the hinge region and the carbohydrate content, the A2m(1) α2 chain differs from the α1 chain in 14 positions; all of these are identical to the A2m(2) α2 chain and confer isotypic (i.e., subclass) specificity on the α2 chains (Figs. 3.9 and 3.10). However, the A2m(2) allotype differs in six positions where A2m(1) and α1 are identical; the first two are just before the hinge region, and the other four are in the last (C_H3) domain. The A2m allotypic character of α2 chains is attributed to several conformational factors in the sequence at positions 211–221, just before the hinge. The isoallotypic determinant shared by α1 chains and the A2m(1) allotype of α2 chains resides in the identity of their C_H3 domains. Thus the A2m(1) α2 chain appears to be a hybrid chain that is identical with α1 in the C_H3 domain and identical with the A2m(2) α2 chain in the C_H1 and C_H2 domains and in the hinge, except for the allotypic determinants arising from two amino acid differences from residues 211–221 (i.e., a Pro ⇆ Ser interchange at position 212 and a Pro ⇆ Arg interchange at position 221). The genetic origin of isotypes, allotypes, and isoallotypes of α chains apparently has involved several events of nonhomologous crossing over and neutral point mutations accumulated late in the evolutionary development of IgA immunoglobulins. Figure 3.10 illustrates a possible sequence of events in the evolution of α-chain C-region genes. The IgA2, A2m(1) molecule, being the only human immunoglobulin that lacks an interchain disulfide bridge between its H and L chains, is likely the newest member of the IgA class. Presumably the *A2m(1)* allele arose from the *A2m(2)* allele by mutation, and a subsequent unequal crossing-over event between the α1 and α2 genes created the hybrid *IgA2, A2m(1)* gene. The

Fig. 3.9 Differences in the amino acid sequences among the C regions of three human IgA α heavy chains of different subclasses and allotypes. The positions of the substitutions shown are in proportion to their sequence numbers. Only those oligosaccharides that differ among the three proteins are shown. (Reproduced with permission from Y. Tsuzukida *et al.*, *Proc. Natl. Acad. Sci.* (*USA*) 76: 1004, 1979)

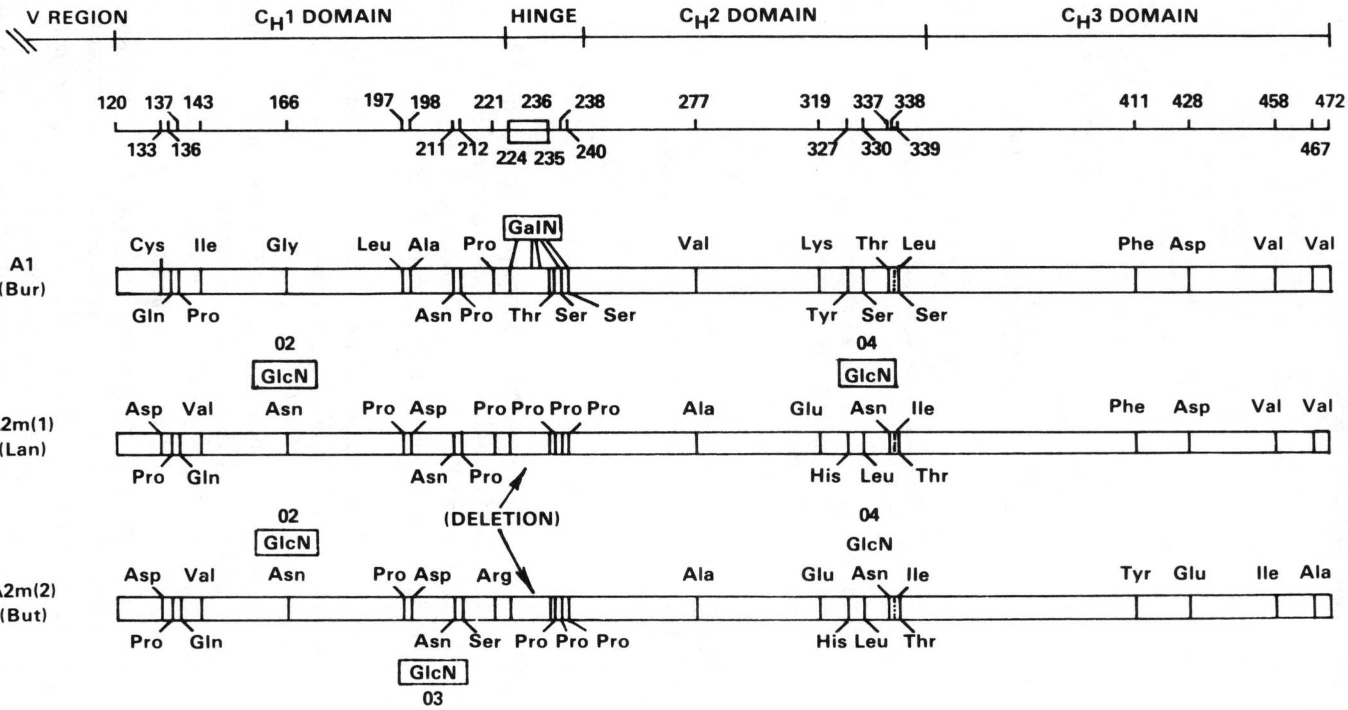

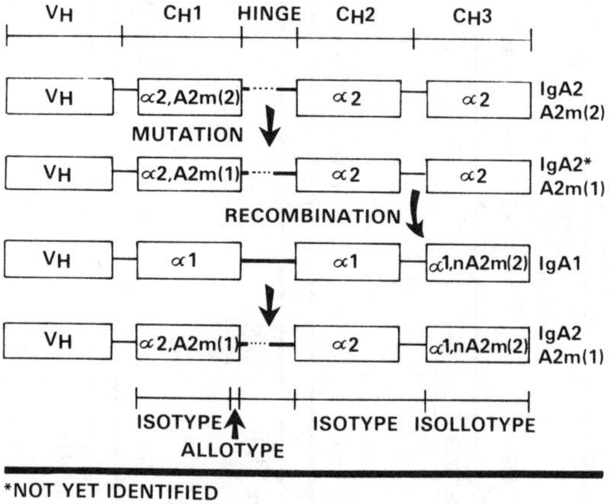

Fig. 3.10 Molecular localization of allotypes, isotypes, and isoallotypes among human α chains. A possible sequence of events regarding the evolution of α-chain C-region genes is indicated.

crossing over may have occurred between C_H2 and C_H3 domains, as H-chain domains are coded for by independent exons in germ cell DNA.

Formal Genetics

The genetically polymorphic antigenic determinants of human immunoglobulins are inherited in three independent linkage groups. One includes the Gm, Mm, and Am markers; the others include the Km and the lambda chain markers, respectively.

Light chains. Three factors have been well characterized in the Km system, namely, Km(1), Km(2), and Km(3) (there is suggestive evidence for a fourth, Km—). Family and population studies show that a single locus controls the expression of all these factors, which are inherited via three codominant alleles, the $Km^{1,2}$, Km^{3}, and the rare Km^{1} alleles. In heterozygous individuals, products of both alleles may be detected. Chemical studies show that a single amino acid difference distinguishes the Km(1), Km(3), and Km(1,2) factors. Therefore, the mode of inheritance of the C region of κ-type L chains is identical to that of other well-studied proteins such as hemoglobin, haptoglobin, and transferrin. Linkage studies with informative families have demonstrated that the *Km* locus is at least 30 crossover units apart from the *Gm* loci. The *Km* locus and *Gm* loci are located on different chromosomes.

α Chains. To date, no genetic marker has been discovered for the IgA1 subclass. Two markers have been described for the IgA2 subclass. They were designated as A2m(1) and A2m(2). Family studies have shown that mating of a homozygous A2m(1) individual to a homozygous A2m(2) individual produces only heterozygous offspring. This observation, plus the fact that no individual negative for both A2m(1) and A2m(2) has been observed in various human populations, indicates that A2m(1) and A2m(2) are complementary alternatives in a bi-allelic genetic system, the summation of their gene frequencies approaching unity. Studies with informative families also show that the Am markers are linked to the Gm markers but not to the Km markers. This finding complements similar results obtained from rabbit and mouse H-chain allotypic markers and indicates that structural genes for the C regions of all immunoglobulin H chains are located on the same chromosome. Population studies show that the gene frequency of A2m(2) is highest in Africa, decreases in Australia and Asia, and is lowest in Europe. If one accepts the concept that humans originated from Africa in evolution, this gene frequency distribution suggests that A2m(2) arose first, and A2m(1) was generated from A2m(2) by mutation. This conclusion is consistent with the one derived from the biochemical data.

γ Chains. Extensive family and population studies for the Gm factors have been conducted in many laboratories, especially those headed by Steinberg, Ropartz, and van Loghem. Such studies show that Gm factors are inherited in groups termed *haplotypes* (formerly, *phenogroups* or *allogroups*) via codominant alleles, in much the same way as Rh antigens of the red blood cells. Population studies show that Gm haplotypes differ markedly in various ethnic groups (Table 3.5). In Caucasians, for example, G1m(1), G1m(2), G1m(17), and G3m(21) are inherited as a haplotype; hence a normal individual is typed either as Gm(1, 2, 17, 21) or Gm(1—, 2—, 17—, 21—). Similarly G1m(4), G3m(5), G3m(11), G3m(13), and G3m(14) are inherited concordantly in Caucasians, so that a normal Caucasian is either positive for all these factors or negative for all. Pedigree studies show that a heterozygous individual, one who is typed as $Gm^{1,2,17,21,4,5,11,13,14}$, has inherited the genes for G1m(1), G1m(2), G1m(17), and G3m(21) from one parent and G1m(4), G3m(5),

Table 3.5 Gm haplotypes commonly present in several races

Race	Haplotype		
	IgG1 factors	IgG3 factors	IgG2 factors
Caucasoid	4	5, 11, 13, 14	23
	4	5, 11, 13, 14	–
	1, 2	21	–
	1, 2, 17	21	–
Negroid	1, 17	5, 11, 13, 14	–
	1, 17	5, 6, 11, 24	–
	1, 17	5, 6, 11, 14	–
	1, 17	5, 13, 15	–
Mongoloid	1, 17	11, 13, 15, 16	–
	1, 4	5, 11, 13, 14	23
	1, 2, 17	21	
Ainu	1, 17	21	–
	1, 17	13, 15, 16	–
	2, 17	21	–
Bushman	1, 17	21	–
	1, 17	5	–
	1, 17	13, 15	–
	1, 17	11, 13, 14	–
Pygmy	1, 17	5, 6, 24	–
	1, 17	5, 13, 14	–
Micronesian	1, 17	21	–
	1, 4	5, 13, 14	–
Melanesian	1, 17	21	23
	1, 2, 17	21	–
	1, 4	5, 13, 14	–
	1, 17	5, 13, 14	23

G3m(11), G3m(13), and G3m(14) from the other. Thus the data obtained in family studies were compatible with the idea that one genetic locus determines the elaboration of all Gm factors. However, studies on isolated myeloma proteins have demonstrated that G1m(1), G1m(4), and G1m(17) activities are restricted to proteins of IgG1 subclass only, whereas G3m(5), G3m(11), G3m(13), G3m(14), and G3m(21) activities are found on proteins of the IgG3 subclass. The specific association of Gm factors with individual IgG subclasses, plus the fact that all four known IgG subclasses coexist simultaneously in all normal individuals, proves that multiple loci control the expression of Gm factors.

The constant regions of $\gamma 1$, $\gamma 2$, $\gamma 3$, and $\gamma 4$ H chains are therefore under the control of four different structural genes. The extremely low incidence of recombination among these genes can best be explained by postulating that they are very closely linked. In studies of rare Gm haplotypes in two Danish families, Kunkel and Natvig demonstrated that crossover may have occurred between G2m(23) (on $\gamma2$ chains) and G3m(21) (on $\gamma3$ chains) in one family, and between G1m(4) (on $\gamma1$ chains) and G3m(21) in another family. These and other studies suggest that the order of three of the four γ-chain C-region genes on the chromosome is $\gamma2$-$\gamma3$-$\gamma1$. Discovery of two hybrid immunoglobulin molecules resembling the "Lepore" type hemoglobin provided additional evidence for this linkage relationship. One of the hybrid molecules is positive for both G1m(4) (located on the C_H1 homology region of $\gamma1$ chains) and G3m(21) (located on the Fc fragment of $\gamma3$ chains); the other has some of the antigenic properties of both IgG4 and IgG2 but not all the antigenic properties of either subclass. Preliminary chemical studies show that the second hybrid molecule has the C_H1 and C_H2 homology region of a $\gamma4$ chain and the C_H3 homology region of a $\gamma2$ chain. These studies indicate that the order of the cistron coding for the C regions of various human IgG subclasses is $\gamma4$-$\gamma2$-$\gamma3$-$\gamma1$ (Fig. 3.11). However, this order of C_H genes has been challenged by Flanagan and Rabbits recently; their study of cosmid clones containing γ, ϵ, and α C-region genes suggests that $\gamma3$ and $\gamma1$ genes are on the 5′ side of $\gamma4$ and $\gamma2$ genes.

In normal adults, the four IgG subclasses are present in unequal proportions. IgG1 comprises approximately 60–70%, and IgG2, IgG3, and IgG4 approximately 14–20%, 4–8%, and 2–6%, respectively, of

Fig. 3.11 Tentative order of the genes for the four subclasses of human γ chains. The two major allelic gene complexes in Caucasians (top two lines) are shown together with one of the Mongoloid allelic gene complexes (bottom line). The latter demonstrates a case of intragenic crossing over.

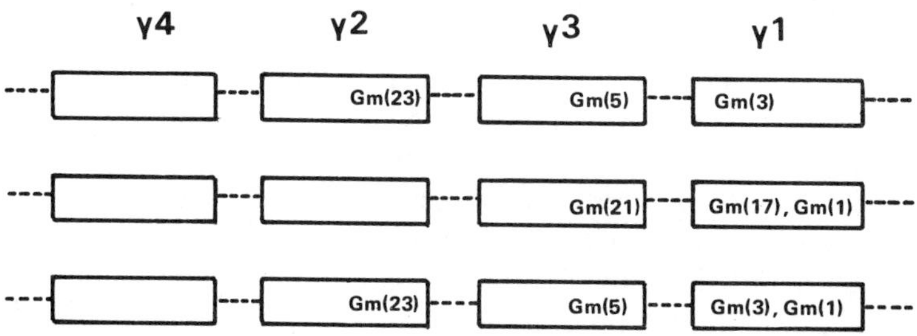

the IgG of normal adults. Although the total level of IgG appears to be rather constant with regard to Gm factors, the levels of the subclasses, at least in Caucasians, appear to be related to the Gm type of the individual. The most pronounced difference, which has been extensively verified, is in IgG3; IgG3 levels are about twice as high in individuals homozygous for G3m(5) as in individuals homozygous for G3m(21); heterozygotes have an intermediate or higher level of IgG3. Similar findings have been observed for IgG2, in that individuals homozygous for G2m(23) have higher levels of IgG2 than individuals homozygous for G2m(23—); again, the level of IgG2 in heterozygotes is usually intermediate between those of the homozygotes. Although no allotypic marker has been described for IgG4, levels of IgG4 appear to be related to the IgG2 allotypic marker G2m(23); G2m(23) homozygotes have the highest levels of IgG4, individuals negative for G2m(23) having the lowest levels and heterozygotes having intermediate levels. These data indicate that the concentrations of all four subclasses of IgG appear to be under genetic control, affected to some extent by the Gm phenotype of the individual. The data indicate that, in Caucasians, individuals homozygous for the allelic factors G1m(4), G2m(23), and G3m(5), on the average, have higher levels of IgG2, 3, and 4 than individuals homozygous for the allelic factors G1m(1), G1m(17), and G3m(21). Furthermore certain Gm factors apparently affect the concentration of other immunoglobulin classes. For example Caucasians homozygous for G1m(4) and G3m(5) have lower serum IgD levels than those homozygous for G1m(1) and G3m(21).

Population distribution of Gm factors. Because different ethnic groups have different Gm haplotypes, anthropologists have been able to use the Gm alleles to study gene drift and migration in ancient times. The credibility of such studies was examined by using the extent of racial admixture of American Negroes. A figure of 30% white admixture was obtained by using the Gm factors as markers. This same figure was obtained in previous studies using hemoglobin S as the marker.

The Gm factors have also been used to examine the extent of Bushman admixture in various Negro tribes in southern Africa. The result, plotted in Figure 3.12, shows that Bushmen probably originated as an isolated group at the southeast corner of Africa.

The results of extensive typing on G1m(1) factor carried out in European populations are shown in Figure 3.13. The frequency of

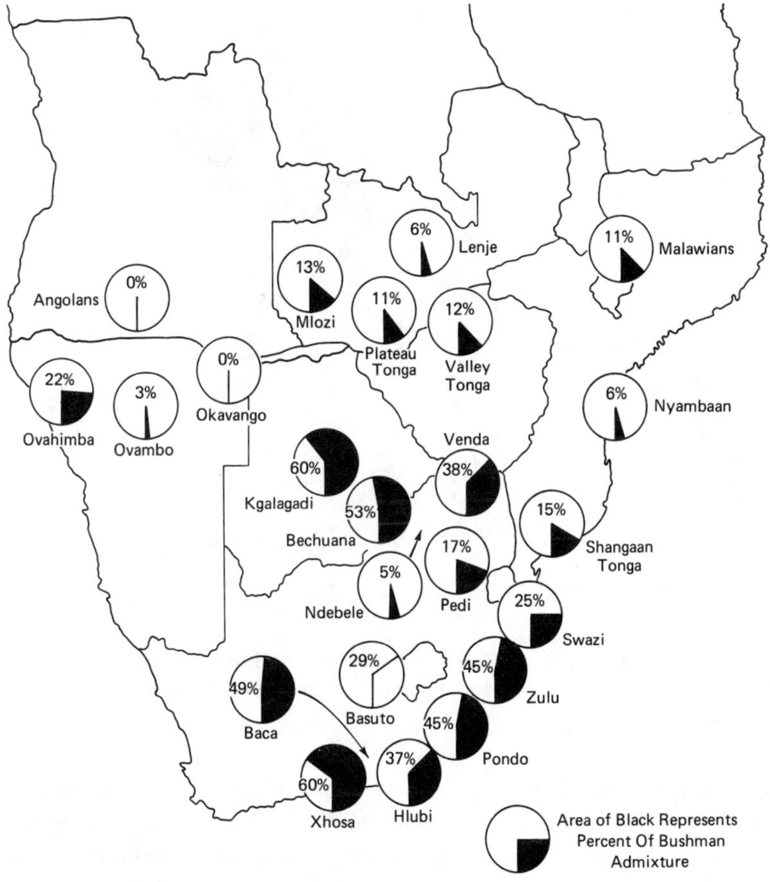

Fig. 3.12 Map showing the amount of Bushmen admixture in various Negro tribes in southern Africa. (From A. G. Steinberg, *Ann. Rev. Genet.* 3: 25, 1969)

the G1m(1) factor is highest at the northeast portion of Europe, with a steady decline toward the south and west.

Studies of G1m(1), G1m(2), G3m(5), and G3m(6) factors in populations in southeast Asia and Australia showed a decline of the $Gm^{1,5}$ allele among the Micronesians, Malays, Thais, Chinese, and Japanese. Asiatic Indians have Gm^{1}, $Gm^{1,2}$, and Gm^{5} alleles, which are characteristic for Caucasians, but not the $Gm^{1,5}$ allele, with the exception of the Oraons, who presumably acquired the $Gm^{1,5}$ allele by racial admixture. Australian Aborigines are 100% G1m(1) positive. The Aborigines of the Western Desert of Australia are univer-

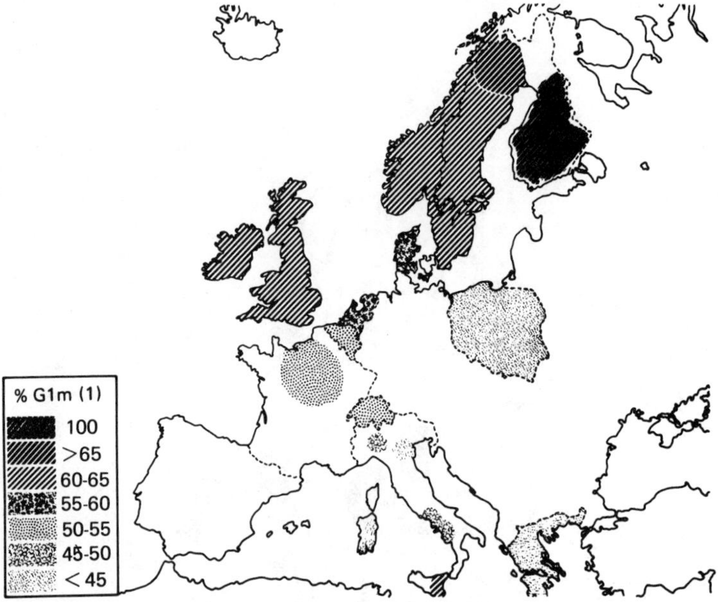

Fig. 3.13 Distribution of G1m(1) phenotype in Europe. The frequency of G1m(1) declines markedly and progressively from north to south and from west to east.

sally negative for G3m(5), but the $Gm^{1,5}$ allele has a frequency of about 17% for those at Kimberly, Australia. The allele carrying the G1m(1) factor in Western Desert Aborigines is established as $Gm^{1,17,21}$, which is common in Caucasians and Mongoloids.

The particular distribution of Gm alleles among different races and the existence of clines within races raise a question concerning the nature of the selective mechanism that maintains them. (There is probably a selective mechanism, for Gm factors are found in other species, for example, chimpanzee.) Since Gm factors are located on the H-chain C regions, which are responsible for many biological activities (e.g., complement fixation, passive cutaneous anaphylaxis, transplacental passage), it is possible that the selective force(s) for C-region genes would be associated with the biological activity rather than with the antibody activity of the immunoglobulin molecules. However, scattered reports suggest that certain antibodies may consist of a single H-chain subclass and of a single L-chain type. (For example, anti-dextran antibodies have been found to be predominantly of γ2 H-chain subclass, and positive association has been found between the $Gm^{1,21}/Gm^{1,21}$ genotype and high respsonsiveness toward flagellin antigen from *S. adelaide.*) Such data, plus the linkage between C- and V-region genes mentioned before, might suggest that the antibody activity may also play a role in the maintenance of Gm alleles.

The Hv(1) allotype. Allotypic variations in the V region of immunoglobulin chains have been extensively studied in rabbits (see p. 84). The first allotypic determinant, designated Hv(1), in the V region of human H chains was identified in our laboratory in 1978. Serological and biochemical studies indicate that Hv(1) is located in the V region of human immunoglobulin H chains of G, M, and A classes. It is inherited as an autosomal dominant trait, with a gene frequency in Caucasians of 0.189 ± 0.023. The polymorphic nature of Hv(1) has also been confirmed in black people: studies on a large black kindred ($n = 80$) have confirmed its autosomal dominant mode of inheritance. The gene frequency in blacks is 0.278 ± 0.067. Hv(1) seems analogous to murine idiotypes of the so-called Idx category, and to rabbit group Aa V_H-regions allotypes, in that it is not restricted to any particular immunoglobulin class. This observation is consistent with our previous finding that the same V-region amino acid sequence is able to associate with C regions of different H-chain classes. However, an important difference between Hv(1) and the idiotypic markers in mice and rabbits is that the former is presumably associated with the framework residues, whereas the latter are associated primarily with the hypervariable regions. Furthermore, the Hv(1) is not linked to the H-chain C-region allotypes (i.e., Gm and Am). Two alternative explanations could be offered for the nonlinkage between Hv(1) and the C_H genetic markers: (1) the V_H genes and C_H genes may be far apart on the same chromosome, and (2) *Hv*(1) may be an allele of a regulatory gene that catalyzes the translocation of V- and C-region genes.

The Isf(1) determinant. An antiglobulin from the serum of a normal individual was found to be specific for an antigenic determinant on certain immunoglobulin molecules. Because the serum was obtained in San Francisco, this determinant was designated as Isf(1). Of 37 IgG myeloma proteins studied, the Isf(1) antigenic determinant was found on 10 IgG1 molecules, but all Bence–Jones proteins as well as IgA myeloma proteins tested were negative. In 48 families with a total of 167 children, polymorphism for the Isf(1) factor was observed. Isf(1) children were not found in families with Isf(1—) × Isf(1—) parents, suggesting a possible dominant or codominant mode of inheritance. No linkage with Gm or Km factors was found. As yet there is no explanation for this observation, but the possibility that a carbohydrate moiety is responsible for the activity has not been ruled out.

Idiotypic Determinants

The term idiotypic determinants was proposed by Oudin in 1966 to define antigenic determinants in a population of antibody molecules that are not observed in other Ig molecules of the same animal or in antibody directed to the same antigen from most other animals of the same species. Initially this term applied only to determinants recognized by antisera prepared in the same species, but now it has also been applied to determinants detected by antisera prepared in different species.

Chemical studies indicate that idiotypic determinants are located in the V regions of H and L chains. These determinants are associated mainly with the hypervariable regions. Studies with inbred strains of mice demonstrate that some strains share cross-reactive idiotypes but others do not. Breeding studies have shown in several mouse strains that cross-reactive idiotypes are inherited in an autosomal codominant fashion. Since these determinants are located at the V regions of Ig polypeptide chains, they are in fact allotypes of the V-region genes. Studies of a number of recombinant mice have allowed the construction of a chromosomal map consisting of three H-chain idiotypic loci (V_H loci) and C_H loci in a linear arrangement. Since idiotypic determinants are usually restricted to antibodies of a single specificity from an individual animal, the establishment of idiotypes as genetic markers of the V region furnishes strong evidence in favor of the germ line hypothesis regarding the genetic control of antibody variability.

ALLOTYPIC MARKERS OF IMMUNOGLOBULINS OF OTHER ANIMALS

Nonhuman Primates

An allotypic marker termed Ma(1) has been described for rhesus monkey immunoglobulins. Immunochemical studies showed that the Ma(1) determinant is located on the Fc fragment of IgG molecules. Family studies indicated that Ma(1) segregates, but the mode of inheritance has not been clearly established because of insufficient data. Evolutionary studies of human immunoglobulins show that some of the human allotypic markers are also present in apes and monkeys. Table 3.6 presents the Gm factors that have been described as occurring in nonhuman primates. The data obtained demonstrate the close correlation between taxonomic classification and the presence of Gm factors. Possibly the Gm factors present on IgG3 H chains evolved before those present on IgG1 H chains. For the factors studied, polymorphism (P) is found only in chimpanzees and orangutans and only for some Gm factors, which in human beings are restricted to IgG3 H chains. No Gm factors are detected in any species lower than the Old World monkeys. The evolutionary origins of the G1m(1) allotypic marker were discussed earlier in this chapter.

Serological tests have also been carried out for A2m(1), A2m(2), Km(1), and Km(3) determinants. Polymorphism is observed for Km(1) factor in chimpanzees, and 25% of chimpanzees tested are Km(1) positive. Caution must be used in the interpretation

Table 3.6 Human immunoglobulin H-chain allotypes in nonhuman primates

Primate		G1m				G3m						A2m	
		1	2	4	17	6	11	13	14	15	16	1	2
Hominoidea (Apes)	Chimpanzee	+	−	−	+	P	P	P	P	P	−	−	−
	Dwarf chimp	+	−	−	+	−	+	−	?	+	−	−	−
	Gorilla	+	−	−	+	−	+	+	−	+	−	−	−
	Orangutan	−	−	−	+	−	P	P	−	+	−	−	−
	Gibbon	+	−	−	+	−	+	−	−	+	−	−	−
Cerco-Pithecoidea (Old World monkeys)	Baboon	−	−	−	+	−	+	−	−	+	−	−	−
	Macacus (5 species)	−	−	−	−	−	+	−	−	+	−	−	−
	Cercocebus fulginosus	−	−	−	−	−	+	−	−	+	−	−	−
	Erythrocebus patas	−	−	−	−	−	−	−	−	−	−	−	−
	Green monkey	−	−	−	−	−	−	−	−	−	−	−	−
Ceboidea (New World monkeys)	*Cebus apella*	−	−	−	−	−	−	−	−	−	−	−	−
Prosimii	Loris	−	−	−	−	−	−	−	−	−	−	−	−

of negative results. Since different antiallotypic antibodies may express slightly different specificity, even toward the same antigenic determinant, negative findings with only one antiserum do not guarantee the absence of the determinant being tested.

Certain human allotypic determinants may be isoallotypic determinants in primates. For example, nearly all the nonhuman primates show a remarkably high serum content of G3m(11). In human beings IgG3 represents less than 8% of the total IgG. Therefore, either G3m(11) occurs in a major IgG subclass in other primates (which is unlikely), or it occurs as an isoallotype in several subclasses.

Primate antisera have also been used to define human IgG antigens. Theoretically rhesus monkeys would not be expected to produce antibodies to the H-chain subclass antigens if they possess the Gm factors and corresponding H-chain subclass determinants of human IgG. This seeming paradox is resolved for antigens of one of the loci by the finding that the IgG3 antigen, as defined by rabbit antisera, was not detectable in rhesus monkey and gibbon sera when a sensitive inhibition assay employing radioactively labeled antigens was used, although IgG3 antigens can be detected in chimpanzees, orangutans, and gorillas. These data suggest that the Gm antigenic determinants in lower primates have evolved separately from the human H-chain subclass determinants.

These data indicate either that gene duplications giving rise to the genes controlling the C regions of the four IgG H-chain subclasses were relatively recent events in evolution, taking place after the divergence of Hominoidae and Cercopithecidea (within 25 million years), or that all primates possess larger numbers of γ-chain C-region genes than those actually expressed and that different sets of $C\gamma$ genes are expressed in different primates.

Mouse

Alloantisera and detection of allotypic markers. Antiallotypic antisera against genetic determinants of mouse immunoglobulins can be produced by immunization of mice of one inbred strain with antibacterial antibodies of another strain coated on bacterial cells or with antibodies directed toward histocompatibility antigens of the host. Antisera prepared in this manner usually contain strong precipitating antibodies. Immunization of rabbits with mouse myeloma proteins can also give rise to antiallotypic antisera. However, the majority of antibodies produced by cross-species immunization react with the appropriate immunoglobulin class derived from all mouse strains, and only a small proportion of antiallotypic antibody may be present. Under such conditions conventional techniques are not sensitive enough to detect the reaction. Satisfactory results have been obtained with a sensitive "inhibition of radioimmune precipitation" procedure. Although similar in principle to the agglutination inhibition assay described earlier, this assay system uses radioactively labeled purified myeloma protein carrying a certain allotypic determinant as antigen and the antiserum monospecific for this determinant. Mixing of the antiserum with mouse sera or immunoglobulins positive for the allotypic marker inhibits the precipitation of the radioactive labeled antigen, whereas sera or immunoglobulins lacking the marker fail to inhibit precipitation of radioactive antigen.

Since mice generally produce good precipitating antibodies, most assays for allotypic markers in mice are based on some form of precipitation reaction. Routine testing is usually performed on Ouchterlony analysis.* Antiglobulins that give precipitation lines against the whole serum from the immunizing strain are then tested against sera from mice of other strains to determine the strain distribution of the allotypic antigenic determinant(s) being detected. To determine which immunoglobulin class carries the allotypic marker or markers being detected, the antiglobulin must also be tested against purified preparations of each immunoglobulin class or against myeloma proteins. Since, however, most immunizations are carried out with partially purified globulins or whole serum from normal or immunized mice, the antisera produced usually contain antibodies directed to more than one immunoglobulin class. Furthermore, these antisera may identify more than one allotypic marker in a single immunoglobulin class. Both factors must be considered when attempting to characterize the specificity of the antibodies present in the antiserum.

Mouse allotypic markers can also be assayed by the hemagglutination inhibition test, in which mouse red blood cells previously coated with immunoglobulins (e.g., antibodies directed against cell membrane "H-2" antigens) are agglutinated by anti-allotypic antibodies. Antiallotypic antisera can cause strong agglutination of red blood cells from old mice of the NZB strain suffering from auto-

* Double diffusion of antigen and antibody on agar gel. A precipitation line forms when the antigen and the antibody directed toward it meet each other at a proper concentration.

immune hemolytic anemia in which antibodies against the cells are present on the red cell surface.

Nomenclature. The terminology of mouse immunoglobulins is confusing, chiefly because similar nomenclature systems have been used for both the immunoglobulin classes and the gene loci that encode the H-chain C regions. A summary of mouse immunoglobulin classes, C_H genes, and their alleles is shown in Table 3.7. For example, when speaking of antibodies or immunoglobulins as proteins, the class name, for example IgM or IgG2a, is generally used. When discussing genetic (allelically determined) differences between immunoglobulins or antibodies within a given class, the genetic or allotypic notation is used. For example, *Ig-1* is the locus for IgG2a, IgG2a in the BALB/c strain is called *Ig-1a,* and IgG2a in C57BL/10 strain is called *Ig-1b.* Attempts to simplify these nomenclatures have met with little or no success as yet. Alternate systems have been suggested, but no agreement on nomenclature for mouse immunoglobulins, especially for allotypes, has been reached.

Genetics and chemistry. The first allotypic marker of mouse immunoglobulin was described by Kelus and Moor-Jankowski. They showed that immunization of one inbred strain of mice (C57BL) with antibody derived from another inbred strain (BALB/c) gives rise to antiallotypic precipitating antiglobulin that reacts with a determinant in the sera of all BALB/c and C3H mice but not with the sera of C57BL mice. In this, as in many similar studies, the antibody was administered coated on bacterial cells (e.g., *Proteus vulgaris*).

Genetic polymorphism has been described for three of the mouse

Table 3.7 Mouse immunoglobulin classes, CH loci, and alleles

Class*	CH locus	Number of alleles known in inbred strains	Number of specificities described
IgG2a (γG)	*Ig-1*	8	11
IgA (γA)	*Ig-2*	5	4
IgG2b (γH)	*Ig-3*	6	7
IgG1 (γF)	*Ig-4*	2	2
IgG3			
IgD	*Ig-5*	2	2
IgM (γM)	*Ig-6*	3	4
IgE	ND†	ND	ND

* Potter–Lieberman class notation given in parentheses.
† ND, not described.

Table 3.8 Allotypic determinants of H chains of mouse IgG2a immunoglobulins (the *Ig-1* locus)

Type strain	Allele	Antigenic determinants*										
BALB/c	*Ig-1*[a]	1(G7)	2(G8)	–	–	–	6	7	8	10(G1)	–	12(G6)
C57BL/10J	*Ig-1*[b]	–	–	–	4	–	–	7	–	–	–	–
DBA/2J	*Ig-1*[c]	–	2(G8)	3	–	–	–	7	–	–	–	–
AKR/J	*Ig-1*[d]	1(G7)	2(G8)	–	–	5	–	7	–	–	–	12(G6)
A/J	*Ig-1*[e]	1(G7)	2(G8)	–	–	5	6	7	8	–	–	12(G6)
CE/J	*Ig-1*[f]	1(G7)	2(G8)	–	–	–	–	–	8	–	11	–
RIII/J	*Ig-1*[g]	–	2(G8)	3	–	–	–	–	–	–	–	–
SEA/Gn	*Ig-1*[h]	1(G7)	2(G8)	–	–	–	6	7	–	10(G1)	–	12(G6)

* Herzenberg's nomenclature is used. Potter–Lieberman nomenclature is given in parentheses.

IgG classes (IgG2a, IgG2b, and IgG1) and for the IgA, -M, and -D molecules. Six gene loci were identified as coding for the genetically determined allotypic antigenic determinants on the C regions of H chains of IgG2a, IgG2b, IgG1, IgA, IgD, and IgM mouse immunoglobulin classes. The gene loci and alleles that control the synthesis of IgG3 have not been identified yet.

More than 10 antigenic determinants have been recognized at the *Ig-1* locus. They are listed numerically in Table 3.8. Each of these determinants was defined by the pattern of inhibition of precipitation exhibited by normal mouse sera of eight allelic types in different antigen–antibody precipitation assays. When a determinant was present in more than one of the allelic types (e.g., a and e), a positive assignment of the determinant to a given immunoglobulin H chain could be made through the availability of myeloma proteins for testing in each of these allelic groups. A given myeloma protein is positive for only one (or a few) of the allotypic determinants that a mouse possesses.

The distribution of *Ig-1* alleles has been extensively studied in inbred mouse strains. Information available to date has been summarized by Herzenberg (Table 3.9).

Five allotypic determinants have been found on the Fc fragment of IgA myeloma proteins. Genetic tests show that these five allotypes are under the control of a single gene, the *Ig-2* locus (Table 3.10). The present status of the allotypic determinants of mouse IgG2b immunoglobulins is shown in Table 3.11. They are under the control of the *Ig-3* locus. Six allotypes have been identified at this locus. The assignment of the *Ig-4* locus responsible for the coding of the C re-

Table 3.9 Distribution of *Ig-1* alleles in inbred mouse strains. The type strain is listed at the top of each column

Ig-1^{a}	*Ig-1*^{b}	*Ig-1*^{c}	*Ig-1*^{d}	*Ig-1*^{e}	*Ig-1*^{f}	*Ig-1*^{g}	*Ig-1*^{h}
BALB/cJ	C57BL/10J	DBA/2J	AKR/J	A/J	CE/J	RIII/J	SEA/Gn
BALB/CGa	BAB/PoHz	DBA/IJ	AL/N	NZB/B1	DE/J	DA/Hu	BDP/J
BUB/Bn	B10.D2(new)Hz	I/Ao		NZO/B1	N/Ao	FZ/Di	BSL/Di
CBA/J	B10.D2(old)Hz	JB/Di		NZW	NH/N	STB/Je	P/J
CH1/Ao	C57BL/H	RF/J					SEC/Gn
C3H/Hz	CWB/Hz	SWR/J					
C3H.SW/Hz	C57BL/Ka						
C57BR/cdJ	C57BL/6J						
C57L/J	H-2H/Go						
C58/J	H-21/Go						
F/Ao	LP/J						
H-2G/Go	SJL/J						
JK/Bi	SM/J						
MA/H	STA/Je						
MA/MyJ	WB/Re						
NZY/B1	WC/Re						
PBR/Ao	WH/Re						
PL/J	WK/Re						
POLY1/Ao	58N/Sn						
POLY2/Ao	101/R1						
PRUNT/Ao							
ST/J							
STR/N							
T6/H							
129/RrGa							

Table 3.10 Allotypes of H chains of mouse IgA molecules (the *Ig-2* locus)*

Type strains	Alleles	Specificities			
BALB/cJ, SEA/Gn	*Ig-2*^{a,b}	–	2(A12)†	3(A13)	4(A14)
C57BL/10J	*Ig-2*^{b}	–	–	–	–
DBA/2J, RIII/J	*Ig-2*^{c,g}	1	–	–	–
AKR/J, A/J	*Ig-2*^{d,e}	–	–	3(A13)	–
CE/J	*Ig-2*^{f}	–	–	–	4(A14)

* Herzenberg's nomenclature is used. Potter–Lieberman nomenclature is given in parentheses.

Table 3.11 Allotypes of H chains of mouse IgG2b molecules (the *Ig-3* locus)*

Type strains	Alleles	Specificities						
BALB/cJ, DBA/2J, SEA/Gn	*Ig-3*^{a,c,h}	1	2(H11)	–	4	7	8	
C57BL/10J	*Ig-3*^{b}	–	–	–	4	7	8	9(H9)
AKR/J	*Ig-3*^{d}	1	–	3	–	7	8	–
A/J	*Ig-3*^{e}	1	–	3	–	7	–	–
CE/J	*Ig-3*^{f}	1	2(H11)	3	4	–	–	–
RIII/J	*Ig-3*^{g}	1	2(H11)	–	4?	–	–	–

* Herzenberg's nomenclature is used. Potter–Lieberman nomenclature is given in parentheses.

Table 3.12 Allotypes of H chains of mouse IgG1 molecules (the *Ig-4* locus)

Type strains		Alleles	Specificities
BALB/c	DBA/2J	*Ig-4*^{a,c,d,e,g,h}	1–
AKR/J	A/J, CE/J		
RIII/J	SEA/Gn		
C57BL/10J		*Ig-4*^{b}	–2

gion of IgG1 H chain was based on differences in electrophoretic mobilities of the Fc fragment of IgG1 immunoglobulins in different mouse strains. Two allotypes have been identified at the *Ig-4* locus (Table 3.12).

The allotypic markers of mouse immunoglobulins are inherited in haplotypes (as are the Gm and Km factors in humans). Therefore, an "allele" is actually defined by the simultaneous occurrence of a certain set of allotypic markers in a given inbred strain of mice; such an allele is exactly homologous to a haplotype (or phenogroup, or allogroup) of the Gm factors found on human immunoglobulins. Close linkage of H-chain C-region genes is the most likely genetic basis for the occurrence of haplotypes, in the case of both the Gm factors and the mouse allotypes. Studies with large numbers of inbred mice using antiallotypic antiglobulins showed no recombination (in more than 3000 crosses) between allotypic determinants of *Ig-1* and *Ig-2* loci, indicating that these two loci are very closely linked. Similar studies have demonstrated close linkage between *Ig-1* and *Ig-3*, as well as between *Ig-1* and *Ig-4* loci. Therefore, in both human and mouse immunoglobulins, H-chain C-region genes are closely linked to one another. The genetic evidence for this statement is confirmed by studies of mouse DNA, which have shown directly that H-chain C-region genes are adjacent to one another on mouse chromosome with a sequence of 5′-μ-δ-γ3-γ1-γ2b-γ2a-ε-α-3′. Furthermore, several laboratories have shown that H-chain C-region allotypic markers are linked to certain idiotypic determinants and that the H-chain structural genes are linked to a gene termed *Pre* that specifies prealbumin polymorphism. The map distance between the *Ig-1* locus and the *Pre* locus is approximately nine crossing-over units.

An allotypic marker in the V region of L chains of mouse immunoglobulins has also been described. The difference was found in strains of AKR/J, C58/J, and RF/J. This marker, termed Ib-peptide marker, is present in approximately 5% of the κ chains and is absent from λ chains. Peptide maps showed that the differences were located in a region of the amino acid sequence near the first half-cystine residue of the V region of the L chains. Breeding studies indicated that the expression of the IB-peptide segregates in a classical Mendelian fashion. A positive correlation was noted between the IB-peptide marker and the expression of a thymocyte cell surface antigen termed Ly-3.1, which is associated with linkage group XI on chromosome 6. The exact meaning of this marker is not clear at present. It may represent an allotype at one of the κ-chain V-region genes or an allotype at a genetic locus regulating the expression of the κ-chain V-region genes.

Rabbit

History and methods of detection. Allotypes of rabbit immunoglobulins were first described in 1956, when Oudin immunized rabbits with purified immunoglobulins (antibodies to bacteria produced in other rabbits) and demonstrated that these immunoglobulins could be immunogenic.

Rabbit alloantisera are usually strongly precipitating antisera. Thus most of the work on rabbit allotypes has been performed with precipitation techniques (either simple precipitation in a liquid phase or precipitation after diffusion in agar gels by Ouchterlony or immunoelectrophoretic* techniques). Ouchterlony analyses of new alloantisera are performed against a panel of sera of known allotype. Examination is carefully made for indication of cross-reactions and for double precipitin bands, which suggest the existence of multiple antibodies to antigens in different immunoglobulin classes. Hemagglutination and hemagglutination inhibition have also been used to provide a convenient assay for work with allotype sera. Agglutination titers with alloantisera to coated red blood cells are usually high, so that high sensitivity can be achieved with this technique.

For the production of monospecific antisera the recipient rabbit should have all the allotypes of the donor except the particular one concerned. Antibody production is best achieved by the injection of immunoglobulins that are antibodies either to protein antigens or to bacterial antigens; these are administered complexed with the respective antigen.

It has also been shown that heterologous antisera, such as goat or chicken anti-rabbit IgG, may contain antibodies to rabbit allotypic determinants. Careful absorption to remove antibodies to the class-specific antigens is necessary to reveal the antiallotype antibodies.

Genetics and chemistry. Following Oudin's first description of rabbit allotypes, at least 23 allotypic determinants have been clearly identified, as shown in Table 3.13. These determinants are controlled by at least eight defined genetic loci, namely, *a*, *b*, *c*, *c′*, *d*, *e*, *f*, and *g*. Locus *a* controls determinants a1, a2, and a3, which are thought to be located on the V region of H chains. Chemical studies showed that the H chains of different a allotypes differ from one another by multiple amino acid residues. Such large-scale amino acid substitutions have never been observed in the allelic forms of any non-immu-

* Electrophoresis of antigen on agar gel, followed by double diffusion of the antigen and the antibodies directed toward it. Precipitation arc forms when the two meet at a proper concentration.

markers, whereas the nature of the other three Ms determinants remains to be specified.

Other Animals

In 1967 Barabas and Kelus described an allotypic determinant associated with rat serum proteins. They immunized one inbred strain of rats (black and white hooded) with serum proteins of another inbred strain (Wistar BB) by repeated intraperitoneal injections. The antiserum they obtained formed a precipitation line with sera of rats of the Wistar strain and another inbred strain (Gowans albino) but not with that of the black and white hooded strain. This determinant was designated as Ra-1. Subsequent studies by others demonstrated that this allotypic specificity was associated with the κ-type L chains of immunoglobulins.

Since 1969 several additional allotypic determinants (RL-1, RL-2, RI-1a, RI-1b, W-1, SD-1, etc.) have been reported by various laboratories. All these (as well as the Ra-1) determinants appear to be related to two allotypic specificities hereafter termed RL-1 and RL-2, since this terminology indicates that the determinants are on related rat L chains. Genetic analysis has established that the RL-1 and RL-2 phenotypes segregate via a Mendelian codominant mode of inheritance and that the gene locus is not linked to the major histocompatibility locus (*Ag-B*) or to three coat color genes (*A*, *C*, and *H*). Chemical studies indicate that the RL-1 and RL-2 antigenic determinants are located on the C region of κ chains. Peptide mapping and partial amino acid sequence analysis has demonstrated multiple amino acid substitutions between the C region of RL-1 κ chain and that of RL-2 κ chain. In this respect the RL allotypes in rats resemble the a and b allotypes in rabbits. The possibility that these allotypes may not be allotypes of structural genes for Ig polypeptide was discussed earlier.

A precipitating antiserum was prepared by immunizing rats of the August inbred strain with a monoclonal IgA protein from a LOW/Wsl strain. This antiserum detected an allotypic determinant designated Iα(1a) on rat α chains. Preliminary results indicate that the presence or absence of Iα(1a) character is determined by a pair of codominant alleles at a single locus. The number of serum IgA molecules carrying the determinant closely approximates 100% in homozygote rats and 50% in heterozygote rats. The genetic locus coding for the Iα(1a) determinant is not linked to that coding for the RL determinants.

The occurrence of allotypes of chicken serum proteins was first reported in 1964. Since then, at least eight allotypic specificities have been clearly defined. These were detected by using precipitating antisera prepared by immunization of chickens of various inbred strains with antigen–antibody complexes (e.g., complexes of *Proteus vulgaris* and anti-*Proteus vulgaris* antibodies) isolated from a different inbred strain of chicken. These allotypic specificities are controlled by three (or more) loci via the autosomal codominant mode of inheritance. At least four* of the above-mentioned specificities, namely, $G1^{a}$ through $G1^{d}$, were found in IgG molecules (thus probably on the γ chain), whereas two others, namely, $M1^{a}$ through $M1^{c}$, were found in IgM molecules (thus probably on the μ chain). Breeding studies have shown that the genetic locus controlling the G1 allotypes is closely linked to that controlling the M1 allotypes. This is consistent with the observations made in humans, rabbits, and mice that genes coding for the C regions of various immunoglobulin H chains are all linked.

One allotypic determinant, called Mc1, has been described in cattle. This determinant is located on the IgM molecules and is transmitted via the autosomal codominant mode of inheritance. Two immunoglobulin allotypic specificities, termed $G1^{a}$ and $G1^{b}$, have been described in pigs. These demonstrate autosomal codominant inheritance and express three phenotypes, G1(a+b−), G1(a+b+), and G1(a−b+).

ALLELIC EXCLUSION, SPECIFIC GENE ACTIVATION, AND DIFFERENTIAL GENE EXPRESSION

At present genetic information on biochemical and functional differentiation of homogeneous cell populations of mammalian organisms is still limited. Evidence from studies of erythrocytes indicates that, in a given individual, all terminally differentiated red blood cells possess the same set of blood group antigens and the same type, or types, of hemoglobin, even when the individual studied is heterozygous. For example, in a heterozygote individual of red cell blood group AB, both alleles are active in the sense that A as well as B antigens are present on every red blood cell. The first exception to the general rule that both alleles of a given locus are active in a given (mammalian) cell was reported by Lyon in 1962. Using the

* Two additional allotypic specificities termed Aa1 and Aa2 have also been found in IgG, but it has not been established whether they are identical to two of the G1 alleles.

sex-linked Xg^a gene as a marker, she demonstrated that only one of the two X chromosomes was active in female cells. The inactivation of the X chromosomes in each cell was random, since females heterozygous for the Xg^a gene carry two populations of red blood cells, one Xg(a+), the other Xg(a−), in approximately equal numbers.

Studies on immunoglobulin allotypic markers have demonstrated that in heterozygotes only one of the two alternative markers is expressed on each mature plasma cell. Because structural genes for immunoglobulins are not sex linked, these genes presented the first example of allelic exclusion of autosomal chromosomes in humans. Furthermore, only one of several classes (and subclasses) of H chains and only one of the two types of L chains are produced by a single cell. Steinberg termed this phenomenon "specific gene activation." Comparison of amino acid sequences of immunoglobulin polypeptide chains in different species furnished evidence suggesting that all species may share a large number of functionally similar genes and that different ones of these genes are expressed in different species. The genetic mechanism for allelic exclusion and specific gene activation has not been fully understood. During ontogenic differentiation of lymphocytes, a somatic translocation brings a particular V-region gene to the neighborhood of a particular C-region gene for the synthesis of a given H or L chain. It is likely that this somatic translocation is also responsible for specific gene activation and allelic exclusion of immunoglobulin gene system. Gene cloning and DNA sequence studies show that nonrearranged *Ig* genes in the germ cells are not expressed. They become expressed only after a V–C translocation that brings a given V to a given C gene, and this V–C translocation generally occurs only on one of the two homologous chromosomes. The rest of this chapter is devoted to detailed description on certain phenomena relevant to specific gene activation and allelic exclusion.

Expression of Immunoglobulin Genes on Individual Molecules

Studies of myeloma proteins of human and mouse show that there is a restriction of the number of allotypic markers on myeloma proteins as compared with those present in the serum of the individual from whom the myeloma protein arises. More than 1000 human myeloma proteins have been analyzed; all have shown such restrictions. For example a person can be positive for the allelic

G3m(21) and G3m(5) IgG3 factors, but IgG3 myeloma protein isolated from his serum can be either G3m(21) or G3m(5), never both.

It has been well established that in an individual heterozygous at both *Gm* and *Km* loci, individual immunoglobulin molecules carry only one type of allotypic determinant on each chain. For example, if a Caucasian carries allotypic specificities G1m(1) and G1m(4) of H chains and Km(1) and Km(3) of L chains, his serum contains molecules of specificities G1m(1)/Km(1), G1m(1)/Km(3), G1m(4)/Km(1), and G1m(4)/Km(3). All molecules are symmetrical in the sense that no single molecule carries one H chain of G1m(1) allotype and the other H chain of G1m(4) allotype. A similar restriction applies to the L chain. There is no restriction, however, on the association of L chains with H chains, and because of this, molecules can arise that are not present in either parent.

Similar situations are found in animal allotypes. In rabbits and mice, with the availability of precipitating antiallotype antisera, the lack of asymmetrical molecules has been clearly demonstrated by precipitation of radioactively labeled IgG with various antiallotype sera. For example, in rabbits, anti-d11 antibodies precipitate molecules carrying d11 only, but not d12, leaving d12 molecules in the supernatant. It has been shown that hybrid molecules can be readily produced *in vitro* by the recombination of artificially split half-molecules; hence, no physicochemical bar exists to the association of different allelic determinants in one molecule.

Expression of Immunoglobulin Genes During Fetal Life

Immunoglobulins are demonstrable in the human fetus, beginning at the 11th week of gestation, and increase steadily up to the time of birth. The human fetus normally produces only small amounts of immunoglobulins, but it has the potential to produce IgG, IgM, and IgA immunoglobulins. Demonstration of the fetal production of IgM and IgA is relatively easy, since these two classes of molecules do not readily cross the placental barrier. The amount of IgA in fetal sera is very small, usually at about 2% of the adult level, and it varies from one serum to another, ranging from nondetectable up to about 30% of adult average in extreme cases. The amount of IgM in cord blood is approximately 10% of the average adult level. However, levels of normal adults have been observed in cases of intrauterine infections, especially in the rubella syndrome.

The study of fetal IgG production is more complicated because the vast majority of the IgG present in the fetal circulation is passively transferred from the mother. Furthermore the transplacental passage of maternal IgG molecules has a selective nature; not all IgG subclasses pass through at equal rates. The G1m(1) molecules produced by heterozygous G1m(1)/G1m(1—) fetuses are capable of passing the placenta and may act as antigens to G1m(1—) mothers. As a result de novo production of anti-G1m(1) antibodies has been observed in a G1m(1—) mother carrying a G1m(1) baby. The ability of the fetus to produce IgG immunoglobulins was first demonstrated by the study of allotypic markers. In Caucasians about 15% of genetically G1m(1) babies have G1m(1—) mothers. G1m(1) activity was clearly demonstrated in the cord sera of about 25% of such babies, indicating the production of IgG1 molecules by the fetus. Comparison of γ chains isolated from pooled fetal sera to those from pooled adult sera shows an extra band unique to the fetal H chains. These data suggest the possible existence of H-chain gene(s) active only during fetal life.

Experiments in animals show that the fetus is capable of mounting an immune response *in utero*. Indeed, shortly before birth, the kinetics of the immune response of the lamb in the uterus are the same as those of adult sheep.

The effects of antiallotype antibodies on fetal and neonatal production of gammaglobulin is another biological facet meriting further exploration. It seems possible that interaction of maternal antibody with fetal cells responsible for synthesis of the gammaglobulins might result in deleterious consequences to the fetus. Fortunately the vast majority of the anti-Gm antibodies formed, at least in human beings, are IgM, which do not cross the placenta. However, perhaps as much as 1% of the anti-Gm antibodies are of the IgG type and can thus cross the placenta. Such IgG antibodies may perhaps be important in the etiology of the disease known as transitional hypogammaglobulinemia of infancy, which is characterized by a marked delay in the onset of antibody formation by the neonate, and hence, by repeated infection during the first 2 years of life. Presumably the mother's IgG anti-Gm antibodies crossing the placenta could bind to the immunologically competent cells in the fetus, hampering mitosis of these cells and thus antibody formation. In mice and rabbits transplacental passage of maternal antibody directed toward genetically determined antigens in fetal immunoglobulins subsequently causes significant depression of production of immunoglobulins bearing the specific determinants in the newborn, but there is a compensatory increase in heterozygotes of immunoglobulin molecules carrying the other determinants. This effect is present in rabbits 3 years after birth. The phenomenon of allotype "suppression" and associated "compensation" is interwoven with those of immunological tolerance, regulation of protein synthesis, and antibody formation.

Cellular Expression of Alleles of the Immunoglobulin Loci

The expression of alleles at immunoglobulin loci in heterozygous animals has invariably shown that in general a single immunoglobulyin-synthesizing cell expresses only one set of immunoglobulin loci, one each of V_H, V_L, C_H, and C_L and only one allele at each particular locus. Since only one allelic type of L or H chain is synthesized per cell, the absence of hybrid immunoglobulin molecules in heterozygous individuals is not surprising. A single cell synthesizes only one L-chain type and one H-chain class, and since L- and H-chain combination has been shown to take place before release of the intact molecule from the cell, one would expect to find only homozygous molecules in the serum of heterozygous individuals. This obviates the necessity of invoking preferential association of chains during synthesis within the cell.

The evidence for allelic exclusion within the immune system comes from three main sources:

1. The restrictions on the genetic markers of myeloma protein in human and mouse as compared with the genotype of the individual in whom the myeloma arises. For example, the myeloma protein of IgG1 subclass in an individual who is heterozygous G1m(1)/G1m(4) will contain either G1m(1) or G1m(4), but not both.

2. Immunofluorescence studies of the distribution of cells bearing or secreting immunoglobulins. These studies have shown that virtually all cells carry or secrete immunoglobulin of only one genetic type.

3. The demonstrations that allelic exclusion also holds for antibody in normal immune responses. Since myeloma cells frequently show chromosomal abnormality, one might doubt the value of extrapolation from myeloma to the normal state. Therefore, this evidence should be stressed. For example, in Rh sensitization in humans, the anti-Rh antibody frequently is predominantly of only one Gm type. This has also been shown to be the case for antibodies to polysac-

charide antigens. For example antibodies to dextran and levan appear to be almost exclusively IgG2, and they are usually restricted to either G2m(23) or G2m(23—) in a heterozygous individual, seldom both. This is also true in animals; in heterozygous mice developing autoimmune hemolytic anemia, individual mice express autoantibodies of either parental type but rarely both parental types.

The Enhancer

Antibody-producing cells do not make complete immunoglobulin chains until after the V-J or V-D-J rearrangements have taken place. This raised a question about the nature of the signal that activates expression of the genes. The DNA sequences that were thought to be involved in initiating transcription of DNA into mRNA are located upstream from the beginning (5′) end of the V region segment. But this 5′ flanking region is not altered by the DNA rearrangements, except that it is brought much closer to the constant region exons and the intron immediately preceding them.

A stretch of DNA designated enhancer has been found recently in between the JH and Cu exons. The translocation of a VH gene along with its 5′ flanking region to the neighborhood of the enhancer markedly increases the transcription of the translocated V-C gene pairs. It appears that the enhancer possesses the ability to activate the expression of linked genes.

It is believed that enhancers also exist in κ and λ chain gene families. Preliminary studies also suggest that enhancers can work in a tissue-specific manner. Thus, the immunoglobulin enhancer works in cells that normally produce immunoglobulins but not in cells that do not.

Allelic Exclusion in Lower Organisms

Allelic exclusion of autosomal genes is not unique to the immunoglobulin system. There are other well-established mechanisms in other species in which a single diploid cell expresses one of two alleles, for example, the *cis*-dominant operator constitutive mutant in bacteria. The phenomenon of specific gene activation and allelic exclusion appears applicable to a wide variety of organisms. Studies on the unicellular organisms *Paramecium aurelia* and *Tetrahymena pyriformis* by Sonneborn, Nanney, and others offer the best example. Such studies have shown that the cilia of these ciliates contain many components; among these, one protein has been isolated and characterized that is uniquely involved in an immobilization response when a cell is exposed to an antiserum directed toward this protein. However, different clones, and different subclones of the same clone respond in different ways to batteries of antisera and can be shown to have different immobilization antigens. When different subclones of a given clone are examined, one finds that a subclone may express any one of several different immobilization antigens. In general most clones have the capacity to produce a dozen or more different immobilization antigens, but ordinarily a particular cell produces only one at a time. Breeding experiments have shown that each antigenic type is associated with a single genetic locus and that the genome of each cell possesses many loci for the antigenic arrays. The variations observed in subclones can hardly be ascribed to mutation, because they are far too common and also because they are to a considerable extent controllable by the environment. In *Tetrahymena* such cellular diversities are not limited to immobilization antigens alone; they are also observed in mating types, esterase types, and types of acid phosphatase.

In other words the subclones of *Tetrahymena* (or *Paramecium*) derived from a single ancestor zygote regularly express considerable variations without changes in genetic constitution of the cells. These variations apparently arise by specific activation (or expression) of members of an array of functionally related genetic loci, or, at least in *Tetrahymena*, alleles of a certain locus.

Evidence for the "Regulatory Gene Control" of Immunoglobulin Synthesis

Very little is known about the genetically controlled regulatory systems that must exist in higher organisms, except that they are probably even more complex than the various well-studied mechanisms that control enzyme synthesis in bacteria or bacteriophage. Nevertheless some findings are difficult to fit into our current concept of the immunoglobulin structural gene linkage groups. These findings might be interpreted as evidence for the existence of regulatory systems controlling immunoglobulin synthesis.

First is the clear-cut evidence that deficiency of an immunoglobulin class or subclass need not be caused by a deletion or malfunctioning of the structural gene involved. Thus a relatively large number of human beings (about 0.2% of Caucasian populations) have

IgA levels that are almost undetectable (less than 4% of the normal value). In some cases this deficiency appears to be genetically controlled, but the mode of inheritance is not clear, and there seem to be several ways in which the defect can be inherited. However, since examples of IgA deficiency that is not genetically linked to the H-chain structural gene group have been reported, the defect could most reasonably be ascribed to the existence of an aberration in a regulatory system or to a defect in secretion rather than of synthesis. Occasional cases of IgA deficiency in conjunction with deletions of a particular chromosome (number 18) have been reported, but most studied cases of the immunoglobulin deficiency have normal chromosome complements. Because there are cases of chromosome-18 abnormality with normal IgA levels, the suggestion of any role of chromosome 18 in immunoglobulin synthesis is still tentative.

Families in which several members have low levels of only one IgG subclass (often IgG3) have also been studied. Again, in most cases, genetic linkage between the H-chain structural gene cluster and the deficiency trait was absent. These studies are consistent with results obtained from amino acid sequence work on the V regions of γ chains in patients suffering from adult-acquired hypogammaglobulinemia. The genetic defect of this disease appears not to be at the level of structural genes. All these results suggest that, whereas a genetic deficiency can be due to the absence of an immunoglobulin structural gene, it is more often due to defects in what might be tentatively identified as something equivalent to regulatory genes.

The second type of evidence suggesting the existence of control mechanisms for immunoglobulin synthesis is not so clear-cut as the first. It includes a group of experiments in which immunoglobulin production can sometimes be induced in unusual ways (in the absence of antigen). These cases apparently demonstrate induction of synthesis of immunoglobulin molecules carrying an allotypic determinant whose presence is not predicted from knowledge of the individual's genotype. Thus cultured lymphocytes from homozygous G1m(4)/G1m(4) humans have occasionally been reported to synthesize G1m(1) molecules after being stimulated with phytohemagglutinin, a plant extract that induces mitosis in lymphocyte cultures.

In certain experiments, transient production of immunoglobulins carrying allotypes that could not be inherited via Mendelian segregation was observed in certain rabbits and mice. Episomal insertion of hypervariable-region genes was postulated as a possible explanation. In other experiments lymphoid cells from rabbits homozygous for an allotypic *a* marker (a1) have been reported to synthesize IgM molecules carrying a3 determinants; this synthesis occurs after a1 recipients are injected with RNA extracts of spleen cells derived from donor a3 rabbits, or after addition of such RNA extracts to cultured a1 lymphoid cells. Most of the IgG (in contrast to IgM) molecules in this anomalous response carry the expected a1 determinants. In this experiment the ability to synthesize specific antibody carrying the donor allotype is apparently transferred, but the mechanism of transfer of information is quite unclear. It has been suggested that the transfer involves either mRNA or a control element that activates hitherto unexpressed a3 genes in the a1 recipients.

Some work on human patients suffering from immune deficiency diseases (Wiskott–Aldrich syndrome) presents what may be a parallel to this experiment. It appears that the capacity of immunoglobulin synthesis in some patients may be restored by administration of "transfer factor," a low-molecular-weight dialyzable extract obtained from human white blood cells. In this case the ability to mount a specific immune response can apparently be transferred.

None of these experiments could be described as clear-cut or easily interpretable, but if they can be confirmed, explanations postulating the existence of regulatory mechanisms for immunoglobulin synthesis will have to be considered. Better results relevant to this subject have come from amino acid sequence studies of immunoglobulin polypeptide chains. In 1974 Wang and Fudenberg reported the finding of a human myeloma L chain that showed more amino acid sequence homology at its V region to that of chicken L chains than to those of other human L chains. This finding was interpreted as evidence for differential gene expression in different species. The authors suggested that all vertebrates may share a large number of similar V-region genes, but different sets of them may be expressed in different species. Subsequent studies of rat κ chains furnished additional evidence to support this finding. The V regions of many (approximately 30%) of the rat κ chains have more amino acid sequence homology to those of the human L chains belonging to the $V_\kappa I$ subgroup (which represents more than 40% of the total κ chains in humans) than to those of the L chains of any other animal species studied, including mouse. These data were substantiated by yet another finding that rabbit H-chain V-region "allotypes" a1, a2, and a3 are isotypes in hare and that one of these allotypes (a1) has been present in all the human plasma (270) tested thus far.

The idea of differential gene expression offers a rational explanation for the complex allotypes that involve large numbers of amino acid substitutions (e.g., the rabbit H-chain V-region allotypes of the *a* locus, the rabbit κ-chain C-region allotypes of the *b* locus, the human Hv(1) allotype, and the rat κ-chain region allotypes of the *RL* locus). Wang has suggested that such allotypes may not be true allotypes of a structural gene but may instead be allotypes of a hypothetical regulatory gene that controls the expression of sets of functionally related structural genes. Figure 3.14 illustrates this concept

Fig. 3.14 A hypothetical explanation of the puzzling rabbit H-chain V-region allotypes. For simplification only a small fraction of the V_H structural genes are shown that represent those coding for H-chain V region carrying a1 and a2 allotypic markers. Those coding for H-chain V regions carrying a3 allotypic marker as well as X and Y markers are omitted. It is suggested that alleles of a hypothetical regulatory gene are responsible for the expression of different sets of V_H genes. As a result multiple amino acid substitutions were observed between H-chain V regions controlled by different alleles. This hypothetical regulatory gene may be functionally in homozygous condition in other species examined thus far, which explains the failure to find such V-region allotypes in H chains of these species. [Modified from A. C. Wang, *Fed. Eur. Biochem. Symp.* 36: 19, 1974. In *Antibody Structure and Molecular Immunology* (J. Gergelz and G. A. Medgyesi, eds.), Amsterdam, New York, North Holland/American Elsevier, 1975.]

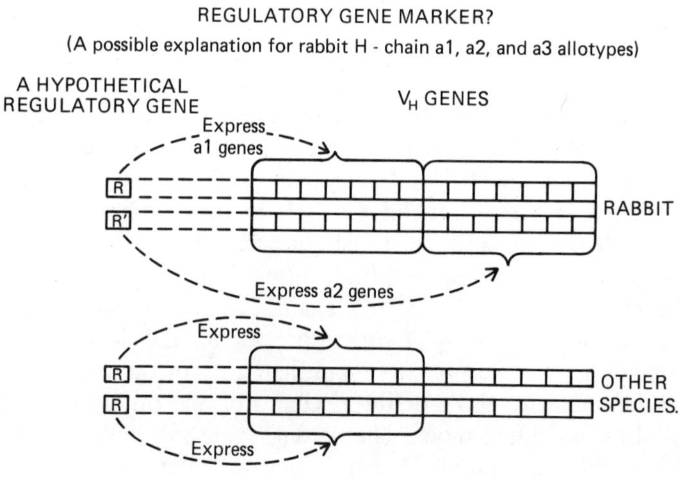

using the rabbit *a* allotypes as an example. It was postulated that all rabbits have genes that code for all the *a* allotypic specificities (i.e., a1, a2, and a3), each of which represents a set of structural genes coding for the H-chain V regions. A control mechanism at a level equivalent to regulatory genes permits them to be expressed so that they mimic a Mendelian pattern of genetic segregation.

RECOMMENDED READINGS

Books

Grubb, R. (1970), *The Genetic Markers of Human Immunoglobulins*. New York: Springer-Verlag.

Klein, J. (1982), *Immunology, The Science of Self–Nonself Discrimination*. New York: Wiley, pp. 156–255.

Nisonoff, A., Hopper, J. E., and Spring, S. B. (1975), *The Antibody Molecule*. New York: Academic Press.

Steinberg, A. G. and Cook, C. E. (1981), *The Distribution of the Human Immunoglobulin Allotypes*. Oxford, New York: Oxford University Press.

Zaleski, M. B., Dubiski, S., Niles, E. G., Cunningham, R. K. (1983), *Immunogenetics*, Boston, London: Pitman.

Reviews

Davis, M. M., Kim, S. K., and Hood, L. (1980), Immunoglobulin class switching: developmentally regulated DNA rearrangements during differentiation. *Cell* 22, 1.

Gally, J. A. and Edelman, G. M. (1972), The genetic control of immunoglobulin synthesis. *Ann. Rev. Genet.* 6, 1.

Giblett, E. R. (1969), The immunoglobulins and their allotypes: Gm and Inv. In *Genetic Markers in Human Blood*. Philadelphia: F. A. Davis Co., p. 3.

Grey, H. M. (1969), Phylogeny of immunoglobulins. *Adv. Immunol.* 10, 51.

Herzenberg, L. A. and Herzenberg, L. A. (1978), Mouse immunoglobulin allotypes: description and special methodology. In D. M. Weir (Ed.), *Handbook of Experimental Immunology* (3rd ed.). Oxford: Blackwell, pp. 12.1–12.23.

Hood, L., Campbell, J. H., and Elgin, S. C. R. (1975), The organization, expression and evolution of antibody genes and other multigene systems. *Ann. Rev. Genet.* 9, 305.

Kindt, T. J. (1975), Rabbit immunoglobulin allotypes: structure, immunology, and genetics. *Adv. Immunol.* 21, 35.

Mage, R., Lieberman, R., Potter, M., and Terry, W. D. (1973), Immunoglobulin allotypes. In M. Sela (Ed.), *The Antigens*. New York: Academic Press, p. 300.

Maki, R., Kearney, J., Paige, C., and Tonegawa, S. (1980), Immunoglobulin gene rearrangement in immature B cells. *Science* 209, 1366.

Natvig, J. B. and Kunkel, H. H. (1973), Immunoglobulins: classes, subclasses, genetic variants, and idiotypes. *Adv. Immunol.* 16, 1.

Pandey, J. P., Whitten, H. D., and Fudenberg, H. H. (1983), Genetics of immunoglobulins. In C. S. David and G. Panayi (Eds.), *Clinical Immunology*. Vol. 1, *Immunogenetics*. Kent, England: Butterworth, in press.

Robertson, M. (1983), What happens when cellular oncogenes collide with Ig genes. *Nature* 302, 474.

Steinberg, A. G. (1969), Globulin polymorphism in man. *Ann. Rev. Genet.* 3, 25.

Steinmetz, M., Winoto, A., Minard, K., and Hood, L. (1982), Clusters of genes encoding mouse transplantation antigens. *Cell* 28, 489.

Strosberg, A. D. (1977), Multiple expression of rabbit allotypes: the tip of the iceberg? *Immunogenetics* 4, 499.

Wang, A. C. (1982), Methods of immune diffusion, immunoelectrophoresis, precipitation, and agglutination. In J. J. Machlonis and G. W. Warr (Eds.), *Antibody as a Tool*. New York: Wiley, pp. 139–161.

Wang, A. C. and Wang, I. Y. (1979), Immunoglobulin structure and genetics. *Rev de' Institut Pasteur* (Lyon) 12, 235.

Wang, A. C., and Wang, I. Y. (1983), Hybrid Ig Molecules: Implication for current theories on immunoglobulin synthesis and evolution. *Surv. Immunol. Res.* 2, 109.

Yarmush, M. L., Sogn, J. A., and Kindt, T. J. (1979), Latent allotypes: a window to a genetic enigma. *Ann. Immunol. (Inst. Pasteur)* 130C, 143.

Zelaschi, D., Newby, C., Parsons, M., van West, B., Sforza-Cavalli, L. L., and Herzenberg, L. A. (1983), Human Ig allotypes: new determinants, new alleles, new methods. *Proc. Natl. Acad. Sci.* 80, 3762.

Articles

Ambrosino, D. M., Siba, G. R., van Loghem, E., and Schiffman, G. (1982), Increased antibody response to immunization with *H. influenzae* Type B and pneumococcal polysaccharide capsules polysaccharide vaccines in adults with the G2m(n) allotype. *Pediatr. Res.* 16, 234A.

Carbonera, A. O., DeMarchi, M., van Loghem, E., and Ansaldi, W. (1983), Gm markers in celiac disease. *Hum. Immunol.* 6, 91.

Clarke, S. H., Claflin, J. L., and Rudikoff, S. (1982), Polymorphisms in immunoglobulin heavy chains suggesting gene conversion. *Proc. Nat. Acad. Sci.* (USA) 79, 3280.

Dreyer, W. J. and Bennett, J. C. (1965), Evidence for somatic rearrangement of immunoglobulin genes coding for variable and constant regions. *Proc. Natl. Acad. Sci.* (*USA*) 73, 3628.

Early, P., Roger, J., Davis, M., Calame, K., Bond, M., Wall, R., and Hood, I. (1980), Two mRNAs can be produced from a single immunoglobulin μ gene by alternative RNA processing pathways. *Cell* 20, 313.

Flanagan, J. G., and Rabbits, T. H. (1982), Arrangement of human immunoglobulin heavy chain constant region genes implies evolutionary duplication of a segment containing, and genes. *Nature* 300, 709.

Hollis, G. F., Hieter, P. A., McBride, O. W., Swan, D., and Leder, P. (1982),

Processed genes: a dispersed human immunoglobulin gene bearing evidence of RNA-type processing. *Nature* 296, 321.

Hozumi, N. and Tonegawa, S. (1976), Evidence for somatic rearrangement of immunoglobulin genes coding for variable and constant regions. *Proc. Natl. Acad. Sci.* (*USA*) 73, 3628.

Kagnoff, M. F. (1982), Two genetic loci control the murine immune response to A-gliadin, a wheat protein that activates coeliac sprue. *Nature* 296, 158.

Marx, J. L. (1983), Immunoglobulin genes have enhancers. *Science* 221, 735.

Mercola, M., Wang, X. F., Olsen, J., and Calame, K. (1983), Transcriptional enhancer elements in the mouse immunoglobulin heavy chain locus. *Science* 221, 663.

Nakao, Y., Matsumoto, H., Miyazaki, T., and Farid, N. R. (1982), IgG heavy chain allotypes (Gm) in atrophic and goitrous thyroiditis. *J. Immunogenet.* 9, 311.

Natvig, J. B., Kunkel, H. G., and Gedde-Dahl, T., Jr. (1967), Genetic studies on the heavy chain subgroups of gamma G globulin. In J. Killander (Ed.), *Gamma Globulins,* Nobel Symposium 3. New York: Wiley Interscience, p. 313.

Pandey, J. P., Ambrosch, F., Fudenberg, H. H., Stanek, G., and Wiedermann, G. (1982), Immunoglobulin allotypes and immune response to meningococcal polysaccharides A and C. *J. Immunogenet.* 9, 25.

Pandey, J. P., Shannon, B. T., Tsang, K. Y., Fudenberg, H. H., and Camblin, J. G. (1982), Heterozygosity at Gm loci associated with humoral immunity to osteosarcoma. *J. Exp. Med.* 155, 1228.

Ravetch, J. B., Kirsch, I. R., and Leder, P. (1980), Evolutionary approach to the question of immunoglobulin heavy chain switching: evidence from cloned human and mouse genes. *Proc. Natl. Acad. Sci.* (*USA*) 77, 6734.

Steinberg, A. G., Milstein, C. P., McLaughlin, C. L., and Solomon, A. (1974), Structure studies of an Inv(1,2) kappa light chain. *Immunogenetics* 1, 108.

Todd, C. W. (1963), Allotypes in rabbit 19S proteins. *Biochem. Biophys. Res. Commun.* 11, 170.

Tsuzkida, Y., Wang, C. C., and Putnam, F. W. (1979), Structure of the A2m(1) allotype of human IgA–a recombinant molecule. *Proc. Natl. Acad. Sci.* (*USA*) 76, 1104.

van Loghem, E. and Litwin, S. D. (1972), Antigenic determinants on immunoglobulins of nonhuman primates. *Trans. Proc.* 4, 129.

Wang, A. C., Mathur, S., Pandey, J., Siegal, F. P., Middaugh, C. R., and Litman, G. W. (1978), Hv(1), a variable region genetic marker of human immunoglobulin heavy chains. *Science* 200, 327.

Wang, A. C., Wang, I. Y., and Fudenberg, H. H. (1977), Immunoglobulin structure and genetics: identity between variable regions of a μ and a $\gamma 2$ chain. *J. Biol. Chem.* 252, 7192.

CHAPTER 4

Antibody Variability

Antibody specificity depends on variation in amino acid sequence. However, because of the heterogeneity intrinsic to the immunoglobulins, chemical differences between antibody preparations were difficult to demonstrate convincingly for a very long time (although many attempts to do so were made, and on occasion small differences in amino acid compositions or peptide map patterns of different antibody preparations were reported). Thus the first demonstrations of the dependence of specificity on sequence were indirect; they were designed to show that denatured antibodies could regain their specificity on being allowed to renature in the absence of antigen. In 1964 Haber and Whitney and Tanford (more than 10 years after the hypothesis that antibody configuration was not dependent on folding around antigen had been proposed) denatured and reduced antibody preparations (specifically, antibody Fab fragments) and found that they partially regained their original activities on refolding and reoxidation. The presence of antigen was not necessary for the reconstitution of antibody activity. It follows that the linear sequence of amino acids in these proteins completely determined the molecules' final structure and activity.

More recently it has been shown that homogeneous antibodies can be raised deliberately by a variety of techniques (Table 4.1), of which the first to be used extensively was the prolonged immunization of rabbits or mice with killed streptococci or pneumococci, whose cell coats contain polysaccharides of known simple structure. By far the most useful of the available techniques, however, is the production of monoclonal antibodies by hybrids between mouse myeloma cells and spleen cells from immunized mice. Hybrid cell lines ("hybridomas") resulting from this procedure can continue to

Table 4.1 Some methods by which antibodies of restricted heterogeneity have been induced

Methods	Antigen	Species
Hyperimmunization to carbohydrate antigens	Streptococcal or pneumococcal vaccine	Rabbit, some mouse strains
Use of monovalent hapten–protein conjugate	DNP-papain*	Rabbit
	DNP-gramicidin-S	Rabbit
	DNP-insulin	Rabbit
Use of cross-reacting antigens	DNP- and TNP-bovine gammaglobulin*	Mouse
	Cross-reacting influenza viruses	Rabbit
Use of limiting antigen doses	Sheep red blood cells	Mouse
Cell transfer at limiting cell dilutions	DNP-bovine gammaglobulin	Mouse
Hybridoma technology	Many different	Mouse, rat, rabbit

* DNP-, TNP-: di- and trinitrophenyl. When antigens modified with these haptens were used, the antibodies investigated were directed against the DNP-hapten.

grow and secrete specific, monoclonal antibody in culture indefinitely (Fig. 4.1).

The amino acid sequence studies that have been done on various homogeneous antibodies show that, first, antibodies of different specificity have different sequences [when either light (L) or heavy (H) chains of several antibodies are compared], and second, that antibodies directed against the same antigen, but quite possibly against different determinants, also often differ in sequence. There are, however, some striking cases of similarity in sequence between antibodies directed against the same antigen, and these are discussed below.

Some myeloma proteins also show specific antibody activity, and these proteins also differ in sequence when they have different specificities and even when they have apparently similar activities (with very few exceptions) (p. 114). There is a great deal of evidence that these proteins are representative antibody molecules, apart from their occasionally demonstrable antibody-like activities; if a pool were made of structural and genetic data on all the human myeloma pro-

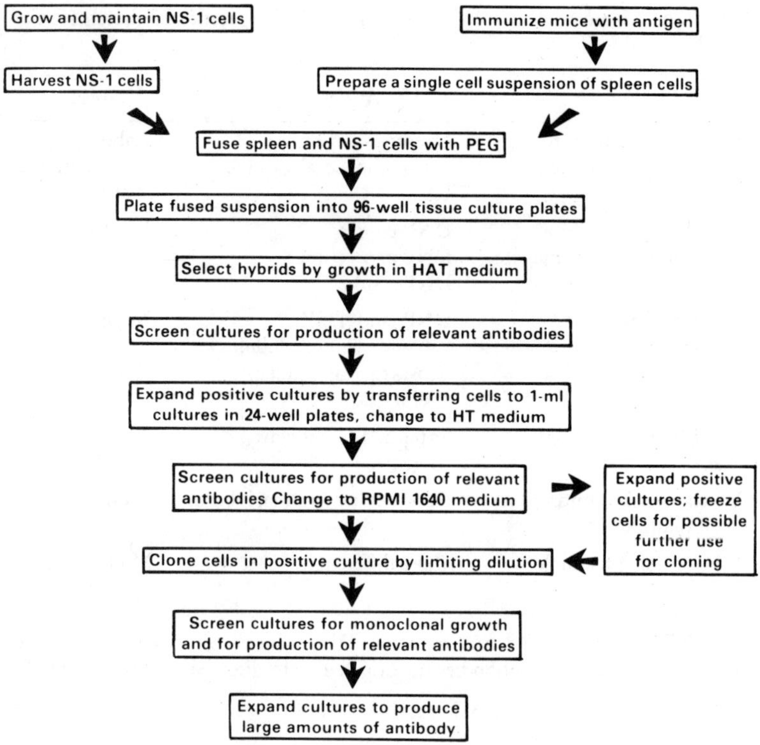

Fig. 4.1 Flowchart indicating steps in the production of hybrid cell lines secreting monoclonal antibody. NS-1 is a mouse myeloma cell line that does not secrete immunoglobulin. It grows in normal tissue culture media (e.g., RPMI 1640) but dies in HAT medium, which contains hypoxanthine, aminopterin, and thymidine. Mouse spleen cells also die in tissue culture, but hybrids between NS-1 and spleen cells, formed in the presence of polyethylene glycol (PEG), survive and grow in HAT medium. (Adapted from B. Mishell and S. M. Shiigi, Eds., *Selected Methods in Cellular Immunology*. New York: W. H. Freeman & Co., 1980, p. 352)

teins studied, the results would be very similar to those obtained by examining a pool of normal immunoglobulins (from many humans).

Any differences that do exist between the normal and the myeloma immunoglobulin pools are probably due not to any intrinsic abnormality of the myeloma proteins, but to the ways in which the condition of myelomatosis is induced. The best example of this is the preponderance of myeloma proteins of the IgA class among myelomas induced in the BALB/c strain of mice by injection of oil into their peritoneal cavities. Presumably most mouse myeloma proteins are IgA because, at the site of induction in the peritoneal cavity, there are more IgA-producing than IgG- or IgM-producing cells (whereas in typical lymphoid tissues, IgG- and IgM-producing cells are more frequent). Peritoneal immunoglobulins may have a different range of specificities from serum immunoglobulins, since the sorts of antigenic stimuli in the two sites are probably different.

Homogeneous antibodies are found, or can be induced, in several situations (Table 4.1). Some naturally occurring homogeneous immunoglobulins have antibody activity; usually these are found in certain disease states and are directed against certain of the patient's own antigens. The *cold agglutinins* are IgM molecules that have antibody activity toward certain red cell antigens (the *I* and *i* antigens), but their ability to agglutinate red cells is manifest only at low temperatures. They almost always have κ L chains; whether this is an indication that only κ chains are associated with antibodies to these red cell antigen specificities or that only κ chains have physical properties enabling such antibody activity to be expressed at low temperatures is not known. The *rheumatoid factors* are almost always IgM molecules that have activity toward IgG determinants and that lead to the formation of immunoglobulin (IgM–IgG) complexes in the serum of some patients with rheumatoid arthritis. *IgG anti-IgG antibodies* are found in other diseases, for example, hypergammaglobulinemic purpura (HGP), in which the presence of complexes is associated with bleeding from capillaries; in several cases these antibodies have also been homogeneous.

PATTERNS OF ANTIBODY SEQUENCE VARIABILITY

The patterns of sequence variability in antibodies and myeloma proteins have become apparent from the immense amount of structural data now available from studies of proteins derived, mostly, from humans, mice, and rabbits. (Several hundred such proteins have been subjected to some structural study.) The type and extent of variability found in immunoglobulin chains can be summarized as follows:

1. Variability is concentrated in the amino-terminal 110–120 residues of the L or H chain (Table 4.2). The concept of a distinct "variable (V) region" has now been so firmly established, not only by the sequence studies mentioned, but also by the examination of peptide maps of perhaps 200 L chains, that any unusual muta-

Table 4.2 V-region sequences of three H chains (one μ and two $\gamma 1$ chains). Variant residues are underlined. Deletions (dashes) have been included to maximize homology between chains. Compare the variability of L-chain sequences (Table 1.1).

			1									10										20		
HE (γG)			PCA	Val	Thr	Leu	Lys	Glu	Asn	Gly	Pro	Thr	Leu	Val	Lys	Pro	Thr	Glu	Thr	Leu	Thr	Leu	Thr	Cys
DAW (γG)			PCA	Val	Thr	Leu	Arg	Glu	Ser	Gly	Pro	Ala	Leu	Val	Arg	Pro	Thr	Gln	Thr	Leu	Thr	Leu	Thr	Cys
OU (γM)			PCA	Val	Thr	Leu	Thr	Glu	Ser	Gly	Pro	Ala	Leu	Val	Lys	Pro	Lys	Gln	Pro	Leu	Thr	Leu	Thr	Cys
							30										40							
Thr	Leu	Ser	Gly	Leu	Ser	Leu	Thr	Thr	Asp	Gly	Val	Ala	Val	Gly	Trp	Ile	Arg	Gln	Gly	Pro	Gly	Arg	Ala	Leu
Thr	Phe	Ser	Gly	Phe	Ser	Leu	Ser	Gly	Glu	Thr	Met	Cys	Val	Ala	Trp	Ile	Arg	Gln	Pro	Pro	Gly	Glu	Ala	Leu
Thr	Phe	Ser	Gly	Phe	Ser	Leu	Ser	Thr	Ser	Arg	Met	Arg	Val	Ser	Trp	Ile	Arg	Arg	Pro	Pro	Gly	Lys	Ala	Leu
		50										60											70	
Glu	Trp	Leu	Ala	Trp	Leu	Leu	Tyr	Trp	Asp	Asp	Asp	Lys	Arg	Phe	—	Ser	Pro	Ser	Leu	Lys	Ser	Arg	Leu	Thr
Glu	Trp	Leu	Ala	Trp	Asp	Ile	Leu	—	Asn	Asp	Asp	Lys	Tyr	Tyr		Gly	Ala	Ser	Leu	Glu	Thr	Arg	Leu	Ala
Glu	Trp	Leu	Ala	—	Arg	Ile	Asx	—	Asx	Asx	Asp	Lys	Phe	Tyr	Trp	Ser	Thr	Ser	Leu	Arg	Thr	Arg	Leu	Ser
							80										90							
Val	Thr	Arg	Asp	Thr	Ser	Lys	Asn	Gln	Val	Val	Leu	Thr	Met	Thr	Asn	Met	Asp	Pro	Val	Asp	Thr	Ala	Thr	Tyr
Val	Ser	Lys	Asp	Thr	Ser	Lys	Asn	Gln	Val	Val	Leu	Ser	Met	Asn	Thr	Val	Gly	Pro	Gly	Asp	Thr	Ala	Thr	Tyr
Ile	Ser	Lys	Asn	Asp	Ser	Lys	Asn	Gln	Val	Val	Leu	Ile	Met	Ile	Asn	Val	Asn	Pro	Val	Asp	Thr	Ala	Thr	Tyr
			100										110											
Tyr	Cys	Val	His	Arg	His	Pro	Arg	Thr	Leu	Ala	Phe	Asp	Val	Trp	Gly	Gln	Gly	Thr	Lys	Val	Ala			
Tyr	Cys	Ala	Arg	Ser	Cys	Gly	Ser	Gln	—	Tyr	Phe	Asp	Tyr	Trp	Gly	Gln	Gly	Ile	Leu	Val	Thr			
Tyr	Cys	Ala	Arg	Val	Val	Asn	Ser	Val	Met															

tions in the constant (C) region of an immunoglobulin chain would now be ascribed to the existence of normal mutational mechanisms (such as have given rise to numerous variants of the human hemoglobin chains) rather than to the action of a special system that produces variants in both the V and C regions of the chain. Occasional C-region variants have been noticed in human λ chains (p. 48).

2. Studies of amino acid sequences of large numbers of V regions showed that each region could be unambiguously classified as belonging to κ, λ, or H chains (Table 4.3). Thus there exists a group of Vκ sequences associated only with κ-chain C regions, a group of Vλ sequences associated with the various λ-chain C regions (p. 48), and a group of V_H (H-chain) V regions associated with any of the 10 or more H-chain C regions.

3. Sequences from each group (κ, λ, or H) of V regions can be further classified into subgroups (Table 4.3). The numbers of κ and λ chain subgroups vary greatly in different species. In most cases this classification is not arbitrary, and can be made by using the criterion that all sequences of the same subgroups show about 80% or greater identity in sequence, whereas proteins of different subgroups are less than 60% identical. A basic sequence can be derived for each subgroup by choosing the most common residue at each position of the proteins belonging to it. In addition antisera have been raised (in rabbits) that clearly distinguish the subgroups of human κ and (most) λ chains. The distinction into subgroups has not therefore been problematic in the great majority of sequences or even partial sequences studied to date. Presumably this is because the evolutionary events leading to the formation of different subgroups are well separated (in evolutionary time) from the events that have generated variability between proteins belonging to the same subgroup.

The evidence from DNA hybridization and sequencing studies (see below) indicates that each subgroup of V-region sequences is coded for by a different gene or set of genes; indeed it is difficult to imagine a mechanism that could produce, from one gene, two sequences that differ by 40% of their residues (as sequences from different human V-region subgroups do). Probably each normal individual has genes coding for members of every L- and H-chain subgroup, since the nonallelic nature of most of the human V-region subgroups has been demonstrated by use of chemical or immunological techniques to identify each of the subgroups in individual normal human sera.

Table 4.3 Subgroups of V regions of κ, λ, and H chains. Nomenclature is that recommended by the World Health Organization. L-chain sequence data are here selected to show the presence of further divisions of each basic sequence (see text). Solid lines indicate identity to basic sequence of corresponding subgroup. Glp indicates pyrrolidone carboxylic acid; Glx indicates glutamic acid or glutamine. Gaps have been arbitrarily inserted in H chains to show their evolutionary relationship to κ chains.

						5					10					15					20
Kappa chains																					
VκI	Basic sequence	Asp	Ile	Gln	Met	Thr	Gln	Ser	Pro	Ser	Ser	Leu	Ser	Ala	Ser	Val	Gly	Asp	Arg	Val	Thr
	Protein Ag	—	—	—	—	—	—	—	—	—	—	—	—	—	—	—	—	—	—	—	—
	Roy	—	—	—	—	—	—	—	—	—	—	—	—	—	—	—	—	—	—	—	—
	Eu	—	—	—	—	—	—	—	—	—	Thr	—	—	—	—	—	—	—	—	—	—
	Mon	—	—	—	—	—	—	—	—	—	Thr	—	—	—	—	—	—	—	—	—	—
	Bel	—	—	—	Leu	—	—	—	—	—	—	—	—	—	—	—	—	—	—	—	—
	Lux	—	—	—	Leu	—	—	—	—	—	Phe	—	—	—	—	—	—	—	—	—	—
VκII	Basic sequence	Glu	Ile	Val	Leu	Thr	Gln	Ser	Pro	Gly	Thr	Leu	Ser	Leu	Ser	Pro	Gly	Glu	Arg	Ala	Thr
	Protein Nig	Lys	—	—	—	—	—	—	—	—	—	—	—	—	—	—	—	—	—	—	—
	Rad	—	—	—	—	—	—	—	—	—	—	—	—	—	—	—	—	Asp	—	—	—
	Cas	—	—	—	—	—	—	—	—	—	—	—	—	—	—	—	—	Asp	—	—	—
	Si	—	—	—	—	—	—	—	—	Ala	—	—	—	—	—	—	—	—	—	—	—
	Win	Asp	—	—	—	—	—	—	—	Ala	—	—	—	—	—	—	—	—	—	—	—
VκIII	Basic sequence	Asp	Ile	Val	Met	Thr	Gln	Ser	Pro	Leu	Ser	Leu	Pro	Val	Thr	Pro	Gly	Glu	Pro	Ala	Ser
	Protein Tew	—	—	—	—	—	—	—	—	—	—	—	—	—	—	—	—	—	—	—	—
	Mil	—	—	—	Leu	—	—	—	—	—	—	—	—	—	—	—	—	—	—	—	—
VκIV	Basic sequence	Asp	Ile	Val	Met	Thr	Gln	Ser	Pro	Asx	Ser	Leu	Ala	Val							
	Protein Len	—	—	—	—	—	—	—	—	Asn	—	—	—	—							
	Jah	—	—	—	—	—	—	—	—	Ala	Thr	—	—	—							
Lambda chains																					
VλI	Basic sequence	Glp	Ser	Val	Leu	Thr	Gln	Pro	Pro	GAP	Ser	Val	Ser	Gly	Ala	Pro	Gly	Gln	Arg	Val	Thr
	Protein HS92	—	—	—	—	—	—	—	—	—	—	—	—	—	—	—	—	—	—		
	HS78	—	—	Ala	—	—	—	—	—	—	—	—	—	—	—	—	—	—	Thr	—	—
	HBJ7	—	—	—	—	—	—	—	—	—	—	Ala	—	—	Thr	—	—	—	Gly	—	—
	HBJ11	—	—	—	—	—	—	—	—	—	—	Ala	—	—	Thr	—	—	—	—		
VλII	Basic sequence	Glp	Ser	Ala	Leu	Thr	Gln	Pro	Ala	GAP	Ser	Val	Ser	Gly	Ser	Pro	Gly	Gln	Ser	Ile	Thr
	Protein HS68	—	—	—	—	—	—	—	—	—	—	—	—	—	—	—	—	—	—	—	—
	HBJ8	—	—	—	—	Ala	—	—	—	—	—	—	—	—	—	—	—	—	—	—	—
	HS86	—	—	Pro	—	Ala	—	—	—	—	—	—	—	—	—	—	—	Glu	—	—	—
VλII	Basic sequence		Tyr	Val	Leu	Thr	Gln	Pro	Pro	GAP	Ser	Val	Ser	Val	Ser	Pro	Gly	Gln	Thr	Ala	Ser
	Protein Kern		—	Ala	—	—	—	—	—	—	—	—	—	—	—	—	—	—	—	—	Val
	X		—	Asp	—	—	—	—	—	—	—	—	—	—	—	—	—	—	—	—	—

Group	Chain	1	2	3	4	5	6	7	8	9	10	11	12	13	14	15	16	17	18	19	20
VλIV	Basic sequence	Glp	Ser	Ala	Leu	Thr	Gln	Pro	Pro	GAP	Ser	Ala	Ser	Gly	Ser	Pro	Gly	Gln	Ser	Val	Thr
	Protein Bo	—	—	—	—	—	—	—	—	—	—	—	—	—	—	—	—	—	—	—	—
	HBJ2	—	—	—	—	—	—	—	—	—	—	—	—	—	—	—	—	—	—	—	—
VλV	Basic sequence		Ser	Glu	Leu	Thr	Gln	Asp	Pro	GAP	Ala	Val	Ser	Val	Ala	Leu	Gly	Gln	Thr	Val	Arg
	Protein Sh	—	—	—	—	—	—	—	—	—	—	—	—	—	—	—	—	—	—	—	—
Heavy chains																					
V_HI	Basic sequence	Glp	Val	Gln	Leu	Val	Glu	Ser	Gly	GAP	Ala	Glu	Val	Lys	Lys	Pro	Gly	Ala	Ser	Val	Lys
	Protein Eu γ1	—	—	—	—	—	Gln	—	—	—	—	—	—	—	—	—	—	Ser	—	—	—
	Ste γ1	—	—	His	—	—	—	Ser	—	—	—	—	—	—	—	—	—	—	—	Met	—
	Ca γ1	—	—	—	—	—	Gln	—	—	—	—	—	—	Arg	—	—	—	—	—		
	Zuc γ3HCD[a]	—	—	—	Val	—	—	—	—	—	—	Asp	Leu	Val	—	—	—	Gly	—	—	—
V_HII	Basic sequence	Glp	Val	Thr	Leu	Arg	Glu	Ser	Gly	GAP	Pro	Ala	Leu	Val	Lys	Pro	Thr	Gln	Thr	Leu	Thr
	Protein Daw γ1	—	—	—	—	—	—	—	—	—	—	—	—	—	Arg	—	—	—	—	—	—
	Cor γ1	—	—	—	—	—	—	—	—	—	—	—	—	—	—	—	—	—	—	—	—
	Ou μ	—	—	—	—	Thr	—	—	—	—	—	—	—	—	—	—	Lys	—	Pro	—	—
	He γ1	—	—	—	—	Lys	—	Asn	—	—	—	Thr	—	—	—	—	—	—	—	—	—
V_HIII	Basic sequence	Glu	Val	Gln	Leu	Val	Glu	Ser	Gly	GAP	Gly	Gly	Leu	Val	Gln	Pro	Gly	Gly	Ser	Leu	Arg
	Vin γ4	—	—	—	—	—	—	—	—	—	—	—	—	Ile	—	—	—	—	—	—	
	Ho γ HCD[a]	Gly	—	Leu	—	—	—	—	—	—	—	Val	Ser	Ile	?	—	—	—	—	—	—
	Wat γ2	—	—	—	—	—	—	—	—	—	—	—	—	—	—	—	—	—	—	—	—
	For α1	—	Ile	—	—	—	—	—	—	—	—	—	—	—	Lys	Gly	—	—	—	—	—
	Low α2	—	—	—	—	—	Glx	—	—	—	—	—	—	—	?	—	—	—	—	—	—
	Jor α1	Glx	—	Glx	—	—	Glx	—	—	—	—	—	—	—	Glx	—	—	—	—	—	—
	Ha α	—	—	—	—	—	—	—	—	—	—	—	—	—	Gly	—	—				
	Wo μ	—	—	—	—	—	—	—	—	—	—	—	—	—	—	—	—	—	—	—	—
	Til μ and γ	—	—	—	—	Leu	—	—	—	—	—	—	—	—	—	—	—	—	—	—	—
	Sha	—	—	—	—	Met	—	—	—	—	—	—	Val	—	Lys	—	—	—	—	—	—
	Wei γ	Asx	—	Glx	—	Met	Glx	—	—	—	—	Ala	—	—	Lys	—	—	Glx	—	—	—
VHIV	Basic Sequence	Glp	Ser	Val	Leu	Asx															
	Dos μ	—	—	—	Ala	—															
	Dau μ	—	—	—	—	—															
	Bus μ	—	—	—	—	—															
	Bal μ	—	—	—	Ala																
	Re μ	—	—	Ala	—																

Within each subgroup (if enough sequences belonging to it are established), further divisions are apparent (Table 4.3). This is particularly clear in mouse and human κ chains. The divisions are based on the presence of a variant from the subgroup basic sequence that is repeated in several members of the subgroup. Repeated variants that occur in proteins belonging to the same subgroup are probably not due to genetic polymorphism, since such variants are also found in myeloma proteins derived from inbred mice (which are presumably identical at all or the great majority of their genetic loci). In addition, the repeated alternatives arginine and glutamine (at position 24 of subgroup $V_{\kappa}I$ of human κ chains) are both probably present in all normal individuals, suggesting that they are derived from two similar but nonallelic genetic loci.

The number of V-region subgroups found in different species varies. Sequence analysis of κ chains from mice has shown that most of the proteins sequenced are not closely related in structure to other κ chains; in this sense (if each chain is a representative of a different subgroup) there are at least 20 and probably as many as 100 subgroups of κ-chain V regions in this species. On the other hand, as discussed below, almost all mouse λ chains studied so far have had very similar sequences. The evolutionary mechanisms leading to such differences are not clear, since in some cases (e.g., mouse, rabbit, and human κ chains) there is no evidence that the V-region subgroups in one species are homologous to those of the other, but in other cases (e.g., the $V_{H}III$ subgroup of various mammalian and avian species) there are clear homologies between subgroups from different species. A likely explanation is that the evolution of V-region genes involves not only point mutation but also frequent gene duplication and elimination; thus sets of closely related genes (such as, perhaps, those coding for V regions carrying the rabbit *a* allotypes discussed on p. 84) may be present in one species but absent in another, but in other cases clear homology may be found between V regions from species not closely related.

4. The wide range of variability seen between different proteins, the concentration of variability in certain regions of the chain, and the stability of sequence of myeloma proteins made by many cells derived from a single tumor all suggest that the variability is not due to errors in misreading a constant genetic message. Experiments designed to test the role of misreading in generating variability have all suggested that it is not an important source of diversity in antibody sequence.

Two questions can be asked about the sequence variability found in the myeloma proteins and antibody molecules studied so far. First, how is the variability in structure related to the primary function of the molecule, that is, its ability to combine with a particular antigen? Second, by what mechanism is the variability generated?

THE ANTIBODY-COMBINING SITE

After Porter's demonstration (in 1959) that binding activity for antigen was located in an antibody's Fab fragments, a number of experiments were done to determine whether the H-chain (Fd) fragment, or the L chain, or both, were the constituents of the Fab responsible for activity. These experiments, which involved separating H and L chains and testing their ability to interact with antigen, were hampered by the difficulties of getting H chains soluble and quite free of L chains; however, in some experiments isolated H chains retained a small percentage of the original activity, and in others activity was lost on chain separation and could be regained only on their recombination, suggesting that both H and L chains contributed to the binding of antigen.

The approximate size of an antibody-combining site was estimated by Kabat and Sela, who showed that, in general, oligosaccharides or oligopeptides of about four or more units' length could inhibit the precipitation reaction between antibody and the homologous polysaccharide or polypeptide. Thus binding of poly-L-alanine by antibodies made against that polymer could be largely inhibited by penta-alanine but not by the tripeptide. The binding-site size for these specific antibody preparations is therefore comparable to that for some enzymes (the active site of lysozyme will accommodate a hexasaccharide), and, in fact, the binding constant for tightly binding inhibitors of lysozyme is comparable to the binding constants of antibodies of normal to high affinity. (Typical values for both constants are around 10^5 M^{-1}.)

The particular portions of the H- and L-chain V regions that are in contact with antigen have been identified by a variety of techniques. Wu and Kabat, comparing the amino acid sequences of a large number of Bence–Jones proteins and myeloma protein L chains, pointed out that three regions of higher than usual variability existed in L-chain V regions; they were located approximately between residues 25 to 35, 50 to 56, and 89 to 96. Similar regions of hypervariability in myeloma or antibody protein H chains were

Table 4.4 The most variable positions in human κ, λ, and H chains. (Numbering is based on sequence of human κ chains.)

Position	Amino acid	Number of occurrences in κ	λ	H chains	Position	Amino acid	Number of occurrences in κ	λ	H chains
28	Ile	0	1		50	Leu	1	0	
	Leu	1	3			Val	1	0	
	Val	2	0			Lys	1	0	
	Met	0	1			Arg	0	1	
	Thr	0	1			Thr	1	0	
	Ser	3	0			Ser	0	1	
	Gly	1	1			Gly	1	2	
	Asp	4	0			Asp	2	0	
						Glu	0	1	
31	Tyr	0	1						
	Phe	0	1		93	Tyr	0	1	
	Lys	1	1			Thr	3	1	
	Thr	2	0			Ser	3	3	
	Ser	3	2			Gly	1	0	
	His	1	0			His	1	0	
	Asn	3	0			Asp	1	0	
						Asn	2	0	
33	Gly			2					
	Thr			3		Glu	1	0	
	Ala			2		Gln	1	0	
	Val			2					
	Ser			1	96	Ile	1	0	
	Tyr			1		Tyr	2	0	
	Phe			1		Phe	2	0	
	Try			1		Pro	1	0	
	Glx			1		Leu	2	0	
						Lys	1	1	
						Arg	1	0	
						Thr	1	0	
						Ser	1	1	
						Asn	0	1	
						Gln	0	1	
						Gln	1	0	
						GAP	0	2	

looked for and found by several groups; the type of variability found (in H and L chains) is shown in Table 4.4 and Figure 4.2.

The evidence that these hypervariable regions form the antibody-combining site comes from comparisons of sequences of myeloma

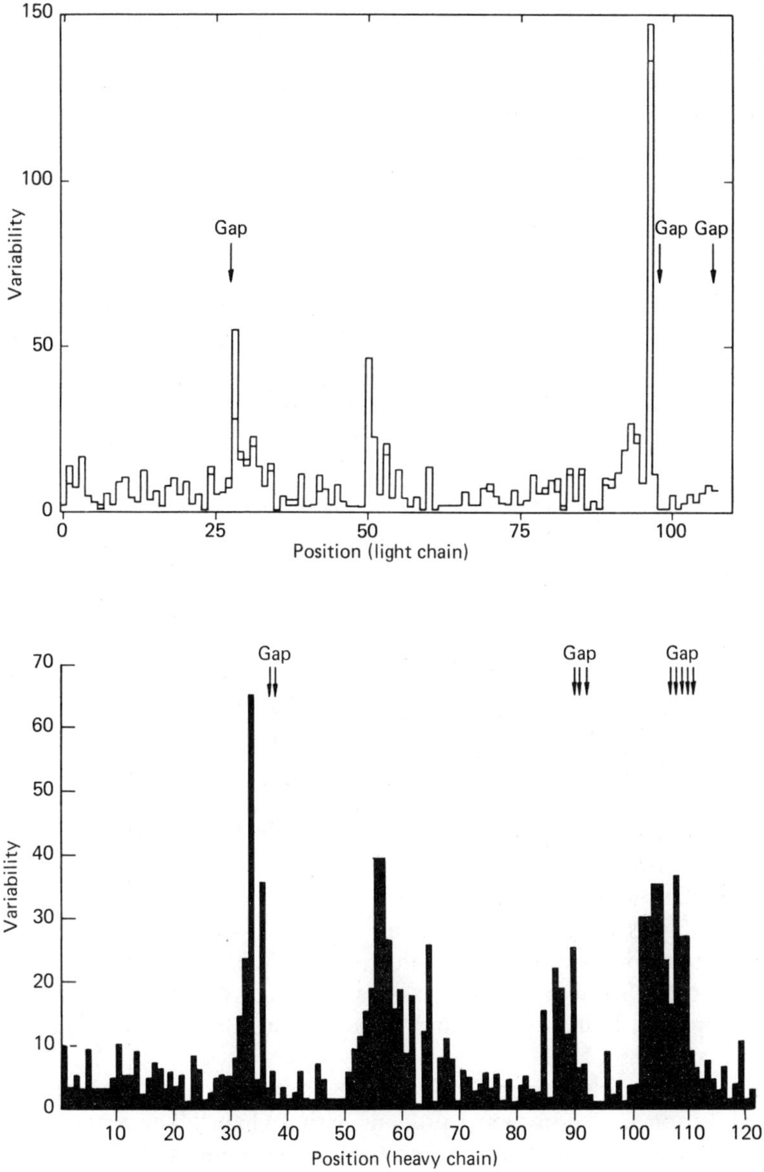

Fig. 4.2 Amino acid sequence variability at different positions of the V regions of immunoglobulin L and H chains. (From T. T. Wu and E. A. Kabat, *J. Exp. Med.* 132: 211, 1970, and J. D. Capra and J. M. Kehoe, *Adv. Immunol.* 20: 1, 1975)

Table 4.5 Some antigens against which various human and mouse myeloma proteins have been shown to possess binding activity. (The differences between species may not be significant, since activity toward the same antigens for human and mouse proteins has not often been examined.)

Human	Mouse
Nitrophenyl haptens	Nitrophenyl haptens
IgG	Dextran
Red cells (in the cold)	Phosphorylcholine
Streptolysin O	Antigens derived from several bacterial strains, including *Pneumococcus*, *Salmonella*, *Escherichia*, *Proteus*, and *Pasteurella*
Heparin	
Lipoprotein	
Cardiolipin	
Phosphorylcholine	

proteins or homogeneous antibodies with known activity, from affinity-labeling experiments, and from X-ray crystallography.

Myeloma Proteins with Antibody Activity

In some cases antibody activities were found more or less by accident; several investigators noticed that, in some patients with myelomatosis, complexes were formed between the myeloma protein produced and some other serum protein (e.g., another immunoglobulin or a serum lipoprotein). More recently Eisen, Potter, and others screened several hundred human and mouse myeloma proteins for their ability to bind a battery of synthetic or naturally occurring antigens, among which are the DNP group, and the polysaccharide components of common infectious bacteria (*Pneumococcus, Streptococcus, Salmonella,* etc.); a small number of myeloma proteins from both human beings and mice react with one or the other of these antigens (Table 4.5).

The most nearly comprehensive purely structural work has been carried out on mouse myeloma proteins with λ chains. The majority of these proteins are IgA, as are most myeloma proteins induced in mice, but they are exceptional in having λ chains, since normal mouse L chains are almost all (more than 95%) of the κ type. In 1966 Appella and Perham, performing limited structural studies on two mouse λ chains derived from different myelomas, found that they were apparently identical. This was later confirmed, and Weigert and Cohn have now studied 18 of these myeloma λ chains; 12 were identical (on the basis of identity of composition of peptides accounting for the entire V-region sequence). The other chains differed only very slightly, at one or two positions, from the basic sequence, and all the differences were located in the hypervariable regions identified by Wu and Kabat at position 25–35, 50–56, and 89–96. The antibody activities associated with some proteins from which the L chains were derived are shown in Table 4.6. These results suggest that mutants in hypervariable regions are selected because they affect the molecule's antibody activity.

Affinity Labeling

A useful technique was developed by Wofsy and Singer for localizing antibody-combining sites to specific regions of the molecule. The method, called affinity labeling, involved the use of an antigen (hapten) analogue chemically modified by the introduction of an additional reactive group, which could form a covalent link with the antibody molecule near the site of binding of the analogue. Some analogues used by Singer are shown in Figure 4.3; analogues of dinitrophenol (DNP) have most often been used to label antibodies raised against a DNP–protein complex.

Affinity labeling of heterogeneous antibody populations has the advantage of giving an "average" result for many different types of molecule but the disadvantage of making further chemical information about the combining site difficult to obtain. Singer tentatively identified the labeled residues of H and L chains in heterogeneous antibody (anti-dinitrophenol) preparations from rabbits, and Franek and Cebra have done the same for antibodies from pigs and guinea pigs. Homogeneous immunoglobulins (myeloma proteins with antibody activity) were studied by Eisen and Metzger, who found that in most cases only one major H- or L-chain sequence became labeled. The results of several of these studies on antibodies and myeloma proteins are collected in Table 4.7, where it can be seen that in all cases, residues in the hypervariable regions were labeled and that different regions have been labeled in different experiments.

X-Ray Crystallography

The three-dimensional structures of several immunoglobulin fragments have recently been determined by using X-ray crystallography

Table 4.6 Tentative partial sequence of the V region of six apparently identical mouse λ chains. (Four other λ chains differed from the basic sequence in the positions shown. Two of the six identical chains are derived from immunoglobulins with binding activity for dextran.)

1, 5, 10, 15, 20, 25

PCA Ala (Val, Val, Thr, Glx, Glx, Ser, Ala, Leu)(Thr, Thr, Ser, Pro, Gly, Thr, Glx, Val, Thr, Leu)(Thr, Cys) Arg Ser Ser

———————————————————————— Asn

————————————————————————

———————————————————————— Thr

———————————————————————— Asn

30, 35, 40, 45, 50

(Thr, Gly, Ala, Val, Thr, Thr, Ser, Asx, Tyr) (Ala, Asx, Trp) (Val, Glx, Glx, Lys, Pro, Asx, His, Leu, Phe) (Thr, Gly, Leu, Ile,

————————————————————————

———————————————————————— Leu

———————— Gly ————————————————

————————————————————————

55, 60, 65, 70, 75

Gly, Gly, Thr, Asx, Asx) Arg (Ala, Pro, Gly, Val, Pro, Ala) Arg (Phe, Ser, Gly, Ser, Leu, Ile, Gly, Asx) Lys (Ala, Ala, Leu)

————————————————————————

————————————————————————

————————————————————————

—— Asn ——————————————————————

80, 85, 90, 95, 100

(Thr, Ile, Thr, Gly, Ala, Glx Thr, Glx, Asx, Glx, Ala, Ile, Tyr) Phe Cys (Ala, Leu, Trp) (Tyr, Ser, Asx, His, Trp) (Val, Phe,

————————————————————————

————————————————————————

————————————————————————

———————————————————— Arg ————

105, 110

Gly, Gly, Gly, Thr) Lys (Leu, Thr, Val, Leu) (Gly, Glx, Pro) Lys

————————————

————————————

————————————

————————————

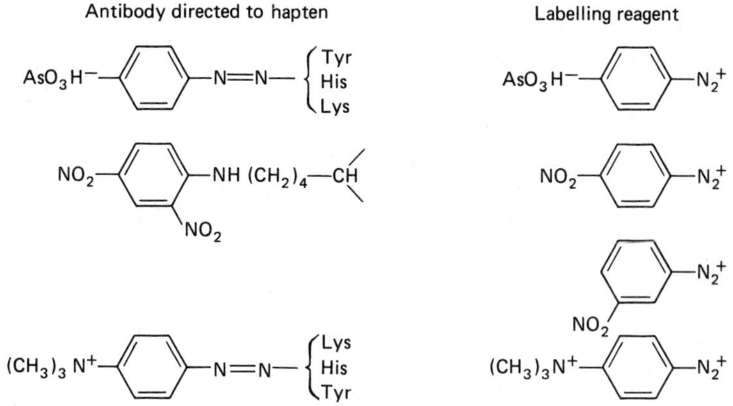

Fig. 4.3 Some affinity-labeling reagents. (From S. J. Singer and R. F. Doolittle, *Science* 153: 13, 1966)

Table 4.7 Results of some affinity-labeling experiments

Protein*	Chain type	Labeled residue and positions	
Mouse anti-DNP antibody	κ	Tyr	86?
Rabbit anti-DNP antibody	κ	Tyr	86?
Pig anti-DNP antibody	λ	Tyr	33
Mouse myeloma protein (anti-DNP)	λ	Tyr	34
Mouse myeloma protein (anti-DNP)	α	Lys	54
Guinea pig anti-DNP antibody	γ	Tyr	32 or 33
Guinea pig anti-ARS antibody	γ	Lys	59
Rabbit anti-NAP antibody	γ	Cys, Ala	97, 98

* DNP, dinitrophenyl; ARS, *p*-azo-benzene arsonate; NAP, *e*-4-azido-2-nitrophenyl.

to a resolution sufficient to allow the peptide chain to be distinguished. The Fab fragments from a mouse phosphorylcholine-binding myeloma protein and from a human myeloma protein later shown to bind vitamin K resemble each other in overall structure: the conclusions drawn from a study of these structures are likely to be general.

The structures show that hypervariable regions from both chains are grouped around a cavity or depression that constitutes the anti-

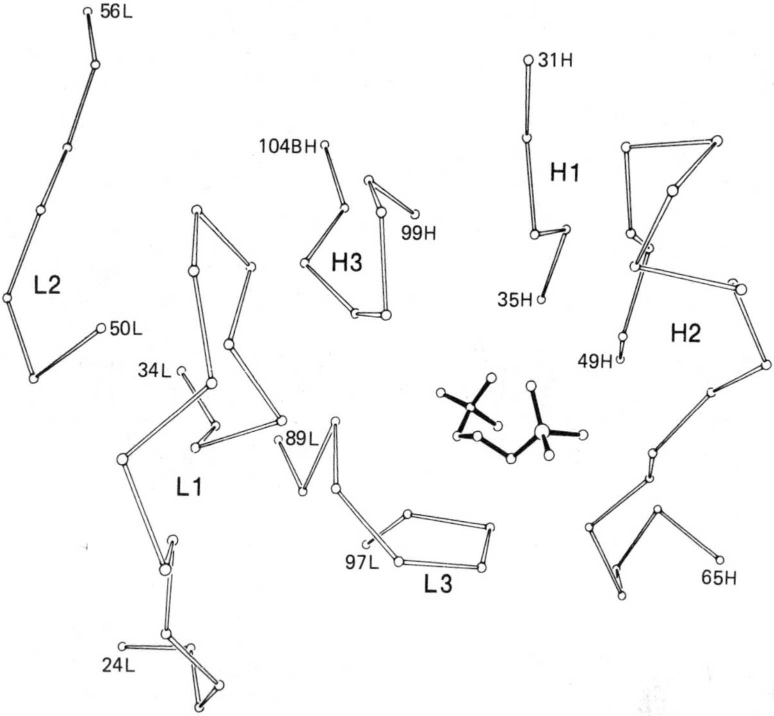

Fig. 4.4 Three-dimensional hypervariable surface of a mouse myeloma protein Fab fragment binding phosphorylcholine. L1, L2, L3, H1, H2, H3 are the L- and H-chain hypervariable regions, respectively. Phosphorylcholine is indicated by black bonds. (From Segal *et al.*, *Progress in Immunology II*, Vol. 1, p. 93, 1974)

gen-binding site (Fig. 4.4). The structure of the Fab fragment does not change significantly on binding antigen, but small conformational changes would not be detected at the resolution used (2–3 Å), or, if they occur, might take place in solution but not in the crystals. The overall structure of the Fab fragments is such that a V domain and a C domain can be seen as separate units joined by short stretches of peptide chain that are exposed to the solvent; the structures of the V and C domains are homologous, and within each domain the structures of the L and H chain are also closely homologous, as would be expected from the amino acid sequence data. The common structure of the domains of each chain—the "immunoglobulin fold"—is illustrated in Figure 4.5, and the tetrahedral

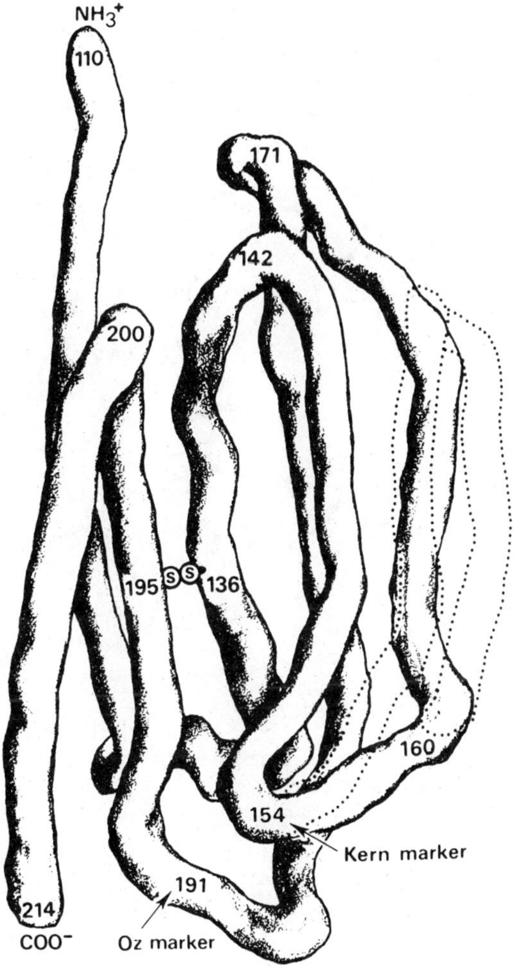

Fig. 4.5 Diagram of the basic "immunoglobulin fold." Solid trace shows the folding of the polypeptide chain in the constant subunits (C_L and C_{H1}). Numbers designate L (λ)-chain residues, beginning at "NH_3^+," which corresponds to residue 110 for the L chain. Broken lines indicate the additional loop of polypeptide chain characteristic of the V_L and V_H subunits. (From Poljak *et al.* *Proc. Natl. Acad. Sci.* (*USA*) 70: 3304, 1973)

Fig. 4.6 Drawing of the α-carbon backbone of the Fab structure of a human myeloma protein. The thicker trace corresponds to the Fd polypeptide, the thinner, to the L chain. H- and L-chain V regions are at the left of the figure, C regions at the right. (From Poljak, *Adv. Immunol.* 21: 1, 1975)

arrangement of these domains to form an Fab fragment, in Figure 4.6.

THE NUMBER OF GENES CODING FOR VARIABLE REGIONS

The great majority of antibody responses are highly heterogeneous: often molecules of all L- and H-chain classes and of a wide range of affinities are produced. In a typical immune response it is not at all obvious whether tens, hundreds, or thousands of different antibody molecular species are produced. In certain cases it has been possible to use the technique of isoelectric focusing, which distinguishes different immunoglobulins on the basis of differences in their isoelectric point with a very high degree of resolution, to estimate the number of different antibodies that mice can make to a particular antigen. Kreth and Williamson used the hapten NIP (nitro-iodo-phenyl) to make such an estimate. In these experiments individual mice immunized with an NIP–protein conjugate made very heterogeneous antibody responses, but the response could be analyzed by transferring spleen cells at limiting dilution to other, irradiated mice; the sera of recipient mice then contained antibodies derived from small numbers of lymphocyte clones. The number of different clones produced by an individual mouse could be counted and the (much larger) total number of different clones produced by all mice of an inbred strain (the antibody repertoire of the strain) estimated. It was shown that individual mice made 30–100 structurally different IgG anti-NIP antibodies in a secondary response, of which almost all were different from anti-NIP antibodies made by other mice. The anti-NIP repertoire of the strain was estimated to be 5000–10,000 antibodies.

Antibody repertoires to other antigens, such as the dinitrophenyl hapten or the influenza virus hemagglutinin coat protein, have been estimated by similar methods (although the less cumbersome hybridoma technology was used to produce the monoclonal antibodies in some cases) and are of the same order of magnitude as the anti-NIP repertoire. Since an individual mouse makes about 100 different anti-hapten or anti-viral protein antibodies, and since it seems quite possible that as many (noncross-reacting) antibodies would be made to the (at least) 10^3–10^4 other antigens that could potentially be used in such an experiment, an individual mouse may be capable of making as many as 10^5–10^6 different antibodies. Another informal argument for this estimate may be made as follows: on immunization the concentration of a specific antibody in an animal's serum may be increased as much as a millionfold, without marked change in total immunoglobulin concentration. If it is assumed that each antibody specificity is present in equal concentration in normal serum, this finding suggests that at least a million different specificities coexist in normal serum.

The postulate of 10^6 different antibody specificities does not in fact require the presence of 10^6 V regions, since, if every L chain could combine (to give a functional molecule) with every H chain, 10^3 V regions for each chain type would suffice. However, in experiments in which separated L and H chains from one antibody preparation have been recombined, either with themselves or with H and L chains from another source, it has been found that there is often preferential recombination of chains derived from the same parent protein; in other words chain combination is not completely random. If, let us say, 10% of H–L chain combinations were useful, then something approaching 10^4 V regions would be needed in total to provide for the synthesis of 10^6 antibodies. Estimates from other sources suggest that is a reasonable figure: for example, the fact that no two identical human L chains have been found (among the more than 100 examined from different myeloma patients) suggests that, probably, several thousand or more different L-chain sequences exist.

Nucleic Acid Hybridization and Sequence Studies

The extent of antibody sequence variability is so great that, if only a few V-region genes are present in the germ line, antibody variability must arise by mutation or recombination during an individual's lifetime. However, there is no evidence from the amino acid sequence data (Table 4.3) that recombination plays any special part in producing antibody variability; if it does, the recombination must be restricted to occurrence between genes coding for variable regions belonging to the same subgroup.

The most direct approach to determining whether somatic mutation is partially responsible for immunoglobulin sequence variability has been to use DNA hybridization cloning and sequencing techniques to isolate and analyze the relevant genes. As discussed in Chapter 3, the V region of a given light chain is coded in the germ line by two separate stretches of DNA: a V gene (with its accompanying leader sequence) and a J gene. The most detailed studies of V- and J-gene number and sequence have been carried out on mouse λ and κ chains by Tonegawa and Leder.

Mouse λ1-producing myelomas, which are probably representative of the λ1 chains from a pool of normal mice, show very limited diversity (Table 4.6); about two thirds of λ1 chains share a single sequence, while the remainder have structures that differ from this basic sequence by a small number (usually one) of point mutations in the hypervariable regions. The total λ1-myeloma-chain repertoire probably consists of at least 15 different sequences. However, nucleic acid hybridization, cloning, and sequence studies show that haploid embryonic DNA contains only one λ1V and one λ1C gene, so that somatic mutation must be responsible for producing the variant sequences.

Since mouse λ1-chain diversity is limited, and these chains constitute only about 5% of total mouse light chains, κ-chain diversity has been equally intensively studied. The exact number of mouse Vκ genes is unknown, but the amino acid sequences of κ-myeloma chains can be divided into a large number (about 50) of different subgroups. Hybridization and restriction enzyme mapping experiments suggest that the number of V genes per subgroup is 5–10, and therefore, that the total number of Vκ genes is of the order of 300. The limited comparisons that have been made of Vκ germ line and protein-encoding genes suggest that somatic mutation contributes to similar extents to Vκ and Vλ variability.

However, two mechanisms not operative in mouse Vλ contribute to Vκ diversity. Mouse κ chains differ from λ chains not only in having many more V genes, but also in having more J genes. Embryonic DNA contains five different Jκ sequences, interspersed with noncoding sequences 250–300 nucleotides long. Amino acid sequence studies of mouse κ chains show that at least four of these five J pieces have been used in different κ chains. Two interesting points emerge from the DNA–amino acid sequence comparisons: first, different Vκ genes can be joined to the same Jκ piece (and vice versa), and second, the exact point at which the V–J recombination is located lies in the third H_V region of the κ gene, but can vary over a stretch of about five nucleotides, so that a single Vκ can be recombined with a single Jκ in several different ways. Table 4.8 illustrates this point by showing how three different sequences, all of which have actually been observed in mouse κ chains, can be generated from one Vκ–Jκ pair.

Table 4.8 Production of different amino acid sequences by V–J recombination at different sites in mouse κ-chain DNA. The amino acid coded by the sequence resulting from recombination at positions 1 to 4 is Trp, Trp, Arg, or Pro, respectively. Complementary DNA sequences (see Fig. 3.4) are boxed. The variable amino acid is residue 96 of the chain.

	Protein sequence
⟶ V sequence	· · · Asp–Pro–TRP–Thr–Phe · · ·
· · · GGATCCTCC CACAGTG · · ·	
1↓	· · · Asp–Pro–TRP–Thr–Phe · · ·
2↓	· · · Asp–Pro–ARG–Thr–Phe · · ·
3↓	· · · Asp–Pro–PRO–Thr–Phe · · ·
4↓	
· · · ACTGTG GTGGACGTTC · · ·	
J sequence ⟶	

Heavy-chain V genes have been less well studied than their light-chain counterparts. Evidence from amino acid sequence and DNA cloning experiments suggests that mouse germ line haploid DNA contains perhaps 100 V_H genes. As noted in Chapter 3, extra heavy-chain diversity arises from the presence of the D (diversity) DNA sequences, which can code for a variable number of amino acids in the chains' third hypervariable region. The total number of D sequences in DNA is not known but is at least 10, and, since it seems likely that 0, 1, or 2 D segments can be placed between V_H and one of several J_H's during heavy-chain V-gene rearrangements, the possibilities for combinatorial variations at the V–J junction are very high. Table 4.9 illustrates the range of diversity encountered in this region of mouse heavy chains.

Searches for Somatic Hypermutation

In the original somatic mutation theories of antibody variability a hypermutation mechanism was postulated to act on V-region genes, perhaps even preferentially on the parts of the gene coding for the hypervariable region of the chains. A number of searches for the existence of such a mechanism, either *in vivo* or in cells growing in culture, have been made. One strategy has been to culture cells producing a myeloma protein and then to select many individual cells from the culture, derive clones from each, and examine their immunoglobulin products by a technique, such as isoelectric focusing, capable of detecting small alterations in amino acid sequence. Milstein, examining 8000 clones derived from a mouse IgG myeloma, found a small number (3–5) that produced altered protein products; a much larger number of mutants that fail to produce or secrete H or L chains, or both, can also be detected in such a system. However, the altered protein products proved on examination to be results of deletion of large parts of the C region, rather than point mutation in the molecule's V region; the positions and sizes of the deletions are illustrated in Figure 4.7. The estimated rate of mutation to such deletion-carrying clones (per gene per cell generation) is of the order of 10^{-5}. The rate of point mutation in the V region is presumably lower, and therefore not enormously different from typical mutation rates (estimated, for example, from studies of the incidence of genetic disorders such as sickle cell disease). Somatic mutation rates in such cases are usually estimated to be less than 10^{-5} per gene per cell division.

Table 4.9 Sequence and length diversity in the third heavy-chain hypervariable region. The sequences are from mouse antibodies to 1,3-dextran (D), galactan (G), or phosphorylcholine (P). (Adapted from Schilling *et al.*, *Nature* 283: 35, 1980)

Activity	V segment	D segment	J segment
D	··· Ala–Arg–Asp–	Tyr–Asp–	Trp–Tyr–Phe ···
D	··· Ala–Arg–Asp–	Val–Asn–	Trp–Tyr–Phe ···
D	··· Ala–Arg–Asp–	Ser–His–	Trp–Tyr–Phe ···
G	··· Ala–Arg–Gly–	Gly–Tyr–Tyr–	Gly–Tyr–Phe ···
P	··· Ala–Arg–Asn–	Tyr–Tyr–Gly–Ser–Thr–	Trp–Tyr–Phe ···
P	··· Ala–Arg–Asp–	Tyr–Tyr–Gly–Asn–Ser–Tyr–	Trp–Tyr–Phe ···
P	··· Thr–Arg–Asp–	Ala–Asp–Tyr–Gly–Asp–Ser–Tyr–Phe–	Gly–Tyr–Phe ···

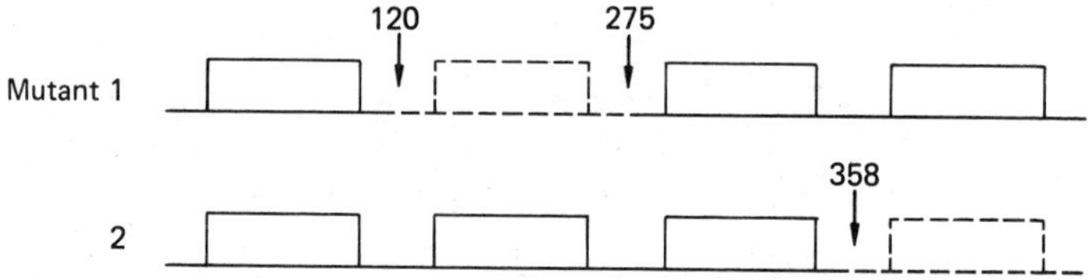

Fig. 4.7 Mutants of mouse myeloma H chains obtained from *in vitro* cultures of myeloma cells. The four disulfide bridge-containing domains of the chain are indicated. The part of the chain that has been deleted is shown dashed, and the residues bounding the deletions are numbered. (Adapted from Milstein *et al.*, *Progress in Immunology II*, Vol. 1, p. 157, 1974)

It might be argued that a myeloma cell, or even a normal IgG-producing cell, had lost a hypermutation mechanism present in precursors of these cells (for example in IgM producers). There is no direct evidence for the existence of such a mechanism, but a possible basis for the recognition of hypervariable regions by a hypothetical mutating enzyme is shown in Figure 4.8, where sequences com-

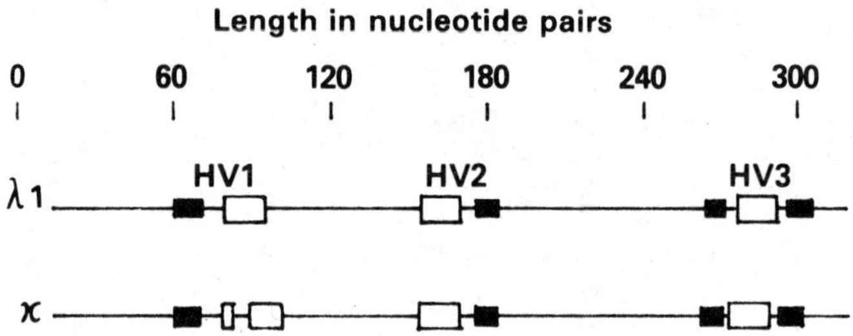

Fig. 4.8 Homologies around and within hypervariable regions (open-boxed areas) of mouse κ and λ genes. There are significant homologies between the three HV regions in the embryonic γ 1 gene, whose sequences are:

HV1: G T T G T A A C A G C C C C A G
HV2: G T G G T A C C A G C A A C C G
HV3: A T G G T A   C A G C A C C C A

(Tonegawa *et al.*, 1978). A gap has been introduced in the HV3 sequence to maximize homology. Differences between sequences are underlined. The sequences are aligned 5 to 3′, and are from the noncoding strand for HV1, but the coding strand for HV2 and 3. There are also homologies between the HV regions of κ-chain genes (Ben-Sasson, 1979). Regions of significant nucleotide homology (> 7 of 8 adjacent bases identical) *between* κ and λ V genes are also indicated (filled boxed areas). A gap has been introduced into the first κ HV region to maximize homology (Wu *et al.*, 1979). (From H. Gutfreund, Ed., *Biochemical Evolution*. Cambridge: Cambridge University Press, 1981, p. 243)

mon to κ or λ hypervariable regions are flanked by other sequences that have been highly conserved during light-chain evolution.

The Possibility of Selection of Somatic Mutants

Since the evidence for the existence of hypermutation mechanisms responsible for somatic mutation is weak, the possibilities of building up a reservoir of mutant cells, by strong selection of mutants that accumulate at a normal rate and by normal mechanisms, have been explored. Cohn and Jerne showed that a mutation rate (one mutant base pair per 10^7 cell divisions) and cell division rate (once every 7 hours) that approached normal values could produce 5000 mutants in 10 days from a total of less than 30 V-region genes; perhaps this number is sufficient to explain the ability of mice to give an immune response 10 days after the start of development of the immune system. Clearly any mutant cell must be very strongly selected for in such a scheme.

The difficulty of such theories is to imagine how such selection could take place. One possibility, suggested by Cohn, is that germ line genes code for V regions that, in particular combinations, form useful antibodies against commonly occurring (i.e., bacterial and viral) antigens. Many of the possible chain combinations will, however, form antibodies unlikely ever to react with an appropriate antigen; these can be selected against by a mechanism leading to the death, after a certain period of grace, of all cells that have not encountered and responded to an appropriate foreign antigen. Mutants arise continually during production of these cells, and those forming combining sites that react with a nonself-antigen are selected; they can in fact undergo further mutation to give better fitting combining sites. Nonself-antigens, according to this scheme, are distinguished from self-antigens by a mechanism such that an antigen must be recognized as foreign by two different cell types (or their antibody products) before a response can be made to it; cells reactive to self-antigens are inactivated, one at a time, as they arise and cannot accumulate to the extent that two such cells will recognize a self-antigen.

An alternative proposal for such a selection mechanism was made by Jerne, who suggested that the selection was for mutants that had lost a postulated antibody activity against self-constituents. The most interesting aspect of this proposal is that it provides a rationale for the genetic linkage between some antibody response genes and histocompatibility antigen genes (Chapter 5). It is suggested that antibody germ line genes in each individual code for antibodies directed against all the histocompatibility antigens of a particular species, and genes coding for antibodies against each individual's antigens can mutate to give altered products of different specificity. Antibodies that have not so mutated are selected against as a consequence of their anti-self activity, and those that have mutated could include many variants active against other antigens.

All the somatic mutants selected for are therefore mutants that have an effect on the molecule's antibody-combining site. This postulate is compatible with the almost total absence of unusual mutants observed in the C regions of immunoglobulin chains. One difficulty of this proposal is that, to obtain mutant chains differing by several point mutations from the product of the parent germ line gene, the affinity of the antibody for a self-constituent should not be lost completely after one mutation has been incorporated but rather should be lost gradually as several mutations are accumulated in the protein's V regions.

It may be possible to show that one or the other (or both) of these hypotheses is unlikely by a study of antibodies against common bacterial and viral antigens and against histocompatibility antigens; if these antibodies are coded for by genes present in the germ line, one would expect to find a high degree of similarity in structure between antibodies with similar specificities but from different individuals. At present the data on such similarities are so limited that no strong conclusion about the specificities of germ line-encoded antibodies can be drawn.

Idiotypes

In most cases studied in detail the antibodies produced by different individuals to the same antigen differ in structure. Isoelectric focusing has been used to show that anti-hapten antibodies produced by different mice are almost always different from one another: for example, among 240 anti-dinitrophenyl antibodies (cross reacting with the trinitrophenyl group) produced by 110 mice in one experiment, fewer than 10 indistinguishable pairs of antibodies were found. Antibodies may also be distinguished by their antigenic characteristics; typically (e.g., in the case of anti-*Proteus*, anti-para-aminobenzoate, or anti-streptococcal carbohydrate antibodies from rabbits) such an antibody carries certain antigenic determinants (individual specificities or "idiotypic" determinants) not detectable on similar antibodies from other animals or on the remaining nonantibody immunoglobulins of the immunized animals. These determinants are located on the V regions of the antibody molecules. Indeed, in some cases, the reactions of anti-idiotype antisera suggest that the idiotype determinants are located near or in the antigen-binding site: thus, if an anti-idiotypic antiserum is prepared against an antibody that has activity toward a hapten X, the presence of X may interfere with the idiotype–anti-idiotype reaction. Such findings of individual differences in antibody structures support the notion that not every V-region sequence is encoded in the germ line (since, if this were the case, many genes would presumably remain totally unused, and there would be no evolutionary selective pressure to maintain them).

There are, however, a number of striking examples of similarity between antibodies from different individuals; the most impressive is the identity in amino acid sequence of λ chains from mouse myelomas, that was previously discussed on p. 114. Several of these myeloma proteins have anti-dextran activity, and the idiotypic determinants carried by two such proteins have been shown to be very similar to each other and to the idiotypic determinants carried by normal anti-dextran antibody raised in the same strain of mice. In another case two mouse κ chains from myeloma proteins with anti-phosphorylcholine activity have been shown to be identical in amino acid sequence, and the idiotypic determinants carried by these proteins were very similar. As a third example, antistreptococcal group A carbodydrate antibodies from many mice hyperimmunized with streptococcal vaccine have been shown to be indistinguishable by isoelectric-focusing and idiotype analysis.

In all these cases the antibodies are directed against antigens that may be components of common pathogenic bacteria. In each case, however, the idiotype similarity of different proteins is confined to antibodies of one specificity derived from a particular strain of mice; antibodies raised against the same antigen in other mouse strains do not carry the same idiotypic determinants. It is therefore possible to use the idiotype characteristics of particular antibodies as genetic markers. In the cases discussed above, the idiotypic markers are encoded by genes closely linked to genes coding for the C regions of mouse immunoglobulin H chains. Thus it is very likely that the expression of the idiotypic markers depends on the structure of the H-chain V region, whose synthesis is controlled by genes linked to the C region (Chapter 3). Several other idiotypic markers have also

Table 4.10 Antibody variable region genes of the mouse

Distance from Ig-1*	Gene	Antibody characteristic determined	Reference
0.4	V_HDEX	Idiotype of antibody (Ab) to dextran, similar to idiotype of J558 myeloma protein	[1]
<1	V_HT15	Idiotype of Ab to phosphorylcholine, similar to TEPC-15 myeloma protein	[2]
<1	V_HARS	Idiotype of Ab to azophenylarsonate hapten	[3]
<1	V_HNP	Fine specificity and isoelectric-focusing pattern of Ab to (4-hydroxy-3-nitrophenyl)-acetyl hapten	[4]
1.4	V_HA5A^{cr}	Idiotype of Ab to streptococcal group A carbohydrate (A-CHO), crossreactive with idiotype of A5A clone	[5]
2.4	V_HInuIdx	Idiotype of Ab to levan, similar to EPC-109 and UPC-61 myeloma protein	[6]
3.2	V_HS117	Idiotype of Ab to A-CHO, similar to S117 myeloma protein	[5]
3.2	V_HA5A	Idiotype of Ab to A-CHO, similar to A5A anti-A-CHO clone from A-strain mice	[7]
4.2	V_HS117^{cr}	Idiotype of Ab to A-CHO, crossreactive with S117 myeloma protein	[5]
5.4	V_HESE	Fine specificity of Ab to sheep red blood cells	[8]
linked†	V_HNBrP	Fine specificity of Ab to (4-hydroxy-5-bromo-3-nitrophenyl)-acetyl hapten	[9]
linked	V_HABA-HOP	Fine specificity of Ab to *p*-azobenzene arsonate hapten	[10]
linked	V_HU10-173	Idiotype of subset (subgroup?) of normal immunoglobulins similar to UPC-10 and MOPC-173 myeloma proteins	[11]
linked	V_HNase-A	Idiotype of Ab to staphylococcal nuclease	[12]
<0.4	V_HNase-S	Idiotype of Ab to staphylococcal nuclease	[12]
linked	V_HKLH	Idiotype of Ab to keyhole limpet hemocyanin	[13]

been shown to be H-chain linked (Table 4.10); in most of these cases (e.g., with the para-aminobenzoate idiotype), the idiotypic determinants on different antibodies are cross-reactive but not identical. In principle it should be possible to construct a genetic map of the H-chain V-region gene complex in this way, but a major difficulty in this proceeding is that, usually, not all mice of a particular strain express the appropriate idiotype on immunization.

Since idiotypes are by definition antigenic, one may ask whether animals can respond to their own idiotypes. Jerne, in answering this question affirmatively, proposed that the immune system forms a "network" in which response to a foreign antigen is followed, and eventually suppressed, by an anti-idiotypic response, which itself induces a further anti-anti-idiotype response, and so on. This idea gains experimental support from the subsequent demonstration of anti-idiotype antibody in mice immunized with pneumococcal vaccine, as well as from the evidence that anti-idiotype antibodies can suppress the synthesis of idiotype-bearing immunoglobulins. However, the complexity of the postulated network is such that it seems unlikely that its manipulation—that is, suppression or enhancement of given immune responses for clinical purposes—is an easily attainable goal.

---

* Percent genetic recombination between Ig-1 allotype locus, a C_H gene, and the listed V_H genes.

† Distance not accurately determined but close linkage shown.

Table provided by Dr. Roy Riblet, The Institute for Cancer Research, Fox Chase, Philadelphia, Pa.

1. Riblet, R. *et al.* (1975), *Eur. J. Immunol.* 5, 778.
2. Lieberman, R. *et al.* (1974), *J. Exp. Med.* 139, 983.
3. Pawlak *et al.* (1973), *J. Exp. Med.* 137, 22.
4. Imanishi, T. and Mäkelä, O. (1974), *J. Exp. Med.* 140, 1498.
5. Zerek, C. *et al.* (1976), *J. Exp. Med.* 144, 1164.
6. Lieberman, R. *et al.* (1975), *J. Exp. Med.* 142, 106.
7. Eichmann, K. and Berek, C. (1973), *Eur. J. Immunol.* 3, 599.
8. McCarthy, M. M., and Dutton, R. W. (1975), *J. Immunol.* 115, 1316.
9. Imanishi, T., and Mäkelä, O. (1975), *J. Exp. Med.* 141, 840.
10. Mäkelä, O. *et al.* (1976), *J. Exp. Med.* 143, 316.
11. Bosma, M. J. *et al.* (1977), *Immunogenetics* 4, 18.
12. Pisetsky, D. *et al.* (1977), *Immunogenetics* 4, 15.
13. Nisonoff, A. (1975), *Immunogenetics* 1, 527.

Books

Cunningham, A. (Ed.) (1976), *The Generation of Antibody Diversity: A New Look*. New York: Academic Press.

Kabat, E. A. (1976), *Structural Concepts in Immunology and Immunochemistry* (2nd ed.). New York: Holt, Rinehart & Winston.

Klein, J. (1982), *Immunology, The Science of Self–Nonself Discrimination* (1st ed.). New York: Wiley, pp. 156–255.

Smith, G. (1973), *The Variation and Adaptive Expression of Antibodies*. Cambridge, Mass.: Harvard University Press.

Tonegawa, S., Sakano, H., Maki, R., Traunecker, A., Heinrich, G., Roeder, W., and Kurosawa, Y. (1980), *Somatic Reorganization of Immunoglobulin Genes During Lymphocyte Differentiation*. Cold Spring Harbor Symposia on Quantitative Biology, Vol. XLV: 839–858.

Reviews

Amzel, L. M. and Poljak, R. J. (1979), Three-dimensional structure of immunoglobulins. *Ann. Rev. Biochem.* 48, 961.

Capra, J. D. and Kehoe, J. M. (1975), Hypervariable region, idiotypy, and antibody combining site. *Adv. Immunol.* 20, 1.

Cohn, M., Langman, R., and Geckeler, W. (1980), Diversity 1980. In M. Fougereau and J. Dausset (Eds.), *Proceedings of the Fourth International Congress of Immunology*, 1980 (*Progress in Immunology* IV). New York: Academic Press, p. 153.

Early, P. and Hood, L. (1981), Allelic exclusion and nonproductive immunoglobulin gene rearrangements. *Cell* 24, 1.

Flanagan, J. G. and Rabbits, T. H. (1982), Arrangement of human Ig heavy-chain constant-region genes implies evolutionary duplication of a segment containing gamma, epsilon, and alpha genes. *Nature* 300, 709.

Gorski, J., Rollini, P., and Mach, B. (1983), Somatic mutations of immunoglobulin variable genes are restricted to the rearranged V gene. *Science* 220, 1179.

Jerne, N. K. (1974), The immune system: a web of V-domains. In *Harvey Society Lectures Series* 70. New York: Academic Press, p. 93.

Leder, P., Max, E. E., and Seidman, J. G. (1980), The organisation of immunoglobulin genes and the origin of their diversity. In M. Fougereau and J. Dausset (Eds.), *Proceedings of the Fourth International Congress of Immunology*, 1980 (*Progress in Immunology* IV). New York: Academic Press, p. 34.

Malcolm, S., Barton, P., Murphy, C., Ferguson-Smith, M. A., Bentley, D. L., and Rabbits, T. H. (1982), Localization of human immunoglobulin kappa light-chain variable-region genes to the short arm of chromosome 2 by *in situ* hybridization. *Proc. Natl. Acad. Sci.* (*USA*) 79, 4957.

Pollack, M. S., Safai, B., Myskowski, P. L., Gold, J. W. M., Landey, J., and Dupont, B. (1983), Frequencies of HLA and Gm immunogenetic markers in Kaposi's sarcoma. *Tissue Antigens* 21, 1.

Propert, D. N., Bernard, C. C. A., and Simons, M. J. (1982), Gm allotypes multiple sclerosis. *J. Immunogenet.* 9, 359.

Tonegawa, S. (1983), Somatic generation of antibody diversity. *Nature* 302, 575.

Tonegawa, S. (1981), Somatic recombination and mosaic structure of immunoglobulin genes. In *Harvey Society Lectures Series* 77. New York: Academic Press, in press.

Wabl, M. and Steinberg, C. (1982), A theory of allelic and isotypic exclusion for Ig genes. *Proc. Natl. Acad. Sci.* (*USA*) 79, 6976.

Wang, A. C. and Wang, I. Y. (1983), Hybrid Ig molecules: implication for current theories on immunoglobulin synthesis and evolution. *Surv. Immunolog. Res.* 2, 109.

Williamson, A. R. (1976), The biological origin of antibody diversity. *Ann. Rev. Biochem.* 45, 467.

Articles

Jerne, N. K. (1970), The somatic generation of immune recognition. *Eur. J. Immunol.* 1, 1.

Köhler, G. and Milstein, C. (1975), Continuous cultures of fused cells secreting antibody of predefined specificity. *Nature* 256, 495.

Rao, D. N., Rudikoff, S., Krütsch, H., and Potter, M. (1979), Structural evidence for independent joining region gene in immunoglobulin heavy chains from anti-galactan myeloma proteins and its potential role in generating diversity in complementarity-determining regions. *Proc. Natl. Acad. Sci.* (*USA*) 76, 2890.

Schilling, J., Clevinger, B., Davie, J. M., and Hood, L. E. (1980), Amino acid sequence of homogeneous antibodies to dextran and DNA rearrangements in heavy chain V-region gene segments. *Nature* 283, 35.

Singer, S. J. and Doolittle, R. F. (1966), Antibody active sites and immunoglobulin molecules. *Science* 153, 13.

Valbueña, D., Marcu, K. B., Weigert, M., and Perry, R. P. (1978), Multiplicity of germ line genes specifying a group of related mouse κ chains with implications for the generation of antibody diversity. *Nature* 270, 780.

Wu, T. T. and Kabat, E. A. (1970), An analysis of the sequences of the variable regions of Bence–Jones proteins and myeloma light chains and their implications for antibody complementarity. *J. Exp. Med.* 132, 211.

CHAPTER 5

Membrane Antigens, Their Genes, and the Immune Response

In mammalian species antibodies are produced by highly specialized plasma cells. These cells are widely distributed in lymphoid organs such as the spleen, lymph nodes, and bone marrow and contain in their differentiated state all of the complex intracellular apparatus associated with immunoglobulin biosynthesis and secretion. The H and L chains of the completed antibody are synthesized along a dense, palisading network of rough endoplasmic reticulum, assembled into a complete immunoglobulin molecule and transported to the exterior of the cell to interact with various other molecules in order to exert their protective role in the economy of the organism.

Specific immune reactions are also mediated by another system, which primarily involves sensitized lymphoid cells. Classically termed cell-mediated or cellular immunity, this system shares many features, such as specificity, anamnestic response, and memory characteristics, with the antibody–plasma cell system. The usual effectors of cellular immunity are small lymphocytes, small rounded cells with disproportionately large nuclei and a narrow rim of cytoplasm. The two major subdivisions of lymphocytes are T lymphocytes, which are primarily involved in cellular immunity, and B lymphocytes, which are partly responsible for the production of circulating antibodies and which differentiate into plasma cells. The T or "thymus-dependent" lymphocytes may also be broadly divided into two main categories, regulatory (helpers and suppressors) and effector cells; however, lymphocytes and other immune-related cells should be regarded as a whole, because they interact extensively.

Cells of different classes carry a variety of distinctive surface molecular markers or receptors, some of which are involved in cell–

cell cooperation and secrete a number of substances important for immune cell–cell interaction. Whereas the substances released by plasma cells are easily categorized as belonging to the various immunoglobulin classes, the substances released by T lymphocytes are extremely diversified. They were initially believed to be the actual effectors of cellular immunity and are released when the cells become activated—usually, but not exclusively, by antigen. Their characterization and activities are discussed later in this chapter.

CELL SURFACE ANTIGENS

B and T lymphocytes carry many different cell-surface markers; some are called differentiation antigens (Ags) because they change as the cell progresses from one differentiation stage to another; many are "alloantigens," because variants of these Ags were distributed in the breeding population as they were codified by allelic genes. Some pertinent well-characterized lymphoid cell surface antigens are listed in Table 5.1.

Table 5.1 Lymphoid cell surface antigens

Antigen	Species	Distribution
I. *Normal Cells*		
MHC (i.e., H-2, HLA)	all	Nucleated cells
TL	all	Thymocytes
NK	all	Natural killer cells
Thy-1	all	Thymocytes, brain cells
Ly	all	B and T cells
Qa	all	T cells
Lyb	all	B cells
Mph-1	all	Macrophages
Lyt	all	T cells
PCA	all	Plasma cells
Ir	all	Lymphoid and others
T	human	T cells
II. *Neoplastic Cells*		
MuLV	mouse	Lymphoid tissue
GCSA	mouse	Lymphoid tissue
GIX	mouse	Thymocytes
X-1	mouse	Lymphoid tissue
TL	mouse	Thymocytes

HISTOCOMPATIBILITY ANTIGENS

A major histocompatibility complex (MHC) of cell surface Ags has been found in all mammalian species studied and is probably present in all vertebrates. Something comparable has been postulated even in the protozoan HLA (histocompatibility-linked antigens). In the mouse, in which the MHC has been studied most thoroughly, it is termed H-2. The MHC is a large cluster of genes coding for cell surface glycoproteins allocated on B and T cells and on macrophages. These glycoproteins exert a profound influence on the interactions of these cells with antigens and with each other. A large part of our knowledge of MHC arose from the study of inbred mouse strains.

The members of these strains (called isogenic or syngeneic) are both genetically uniform among themselves and homozygous at all genetic loci, whereas animals of different strains are allogeneic. Tissue grafts between syngeneic mice succeed, whereas those between allogeneic individuals fail because of incompatibility at the alloantigen level on the cells of the grafted tissue.

The H-2 System in Mice

The MHC was discovered by Gorer in the 1930s while he was studying mouse inbred strains. Originally he attributed the rapid rejection of the test graft to a gene, termed *H-2*, which coded for a "histocompatibility" antigen (termed II). This gene was subsequently shown by Snell (1980) to be a cluster of several genes, called the H-2 complex. This complex is a highly polymorphic genetic system located on chromosome 17 and consisting of regions, subregions, and loci. The complex encompasses the genes controlling the production and/or the expression of glycoproteins embedded in the cell membrane. Several mechanisms for the generation of this extensive polymorphism have been suggested (Weiss *et al.*, 1983), the most recent one being gene conversion. Gene conversion is a meiotic process of directed change in which one allele directs the conversion of a partner allele to its own form. Thus gene conversion could be responsible for the introduction of new sequences in a given histocompatibility gene and hence for the generation of a high number of polymorphic alleles at the H-2 or HLA loci.

The classical organization of the H-2 complex is shown in Table 5.2 together with its division into regions and subregions. The control of different traits is attributed to each region or subregion. Actu-

Table 5.2 The traditional version of the H-2 complex (Reprinted with permission from J. Klein, *Immunology: The Science of Self-Nonself Discrimination*, 1st ed. New York: J. Wiley & Sons, 1982.)

	K	*I*					*S*	*D*
Subregion		*I-A*	*I-B*	*I-J*	*I-E*	*I-C*		
Molecular product	K	Aα, Aβ, Eβ	?	?	Eα	?	SS, Slp	D, L
Trait (locus)								
Serologically detectable (H-2), Ia, SS, Slp)	++++	++++	−	+	++	−	++++	++++
Rejection of allografts (H)	++++	+++	−	−	++	?	−	++++
Cell-mediated lympholysis (H)	++++	+++	−	−	+++	?	−	++++
Mixed lymphocyte reaction (*Lad*)	++	++++	−	+	++	+	−	++
Control of immune response								
(*Ir*)	+	++++	++	−	+++	−	−	−
(*Is*)	−	++++	−	−	−	++++	−	−
T-Cell marker	−	−	−	++++	−	−	−	−
Control of T–B cell collaboration	−	++++	?	?	+++	?	−	−
Restriction of T_c-cell specificity	++++	−	−	−	−	−	−	++++
Restriction of T_h-cell specificity	−	++++	?	?	+++	?	−	−
Control of complement activity	−	−	−	−	−	−	++++	−

ally, each trait is presumably governed by a single locus, but the reciprocal arrangement of various loci into the regions is unknown; what is known is that recombinants were never obtained by separating the different individual loci. The various genes in the H-2 region are indicated by superscripts designating various antigenic sites or structural determinants that are termed the haplotype. For instance, the BALB/c strain, which has an H-2d haplotype, had the *d* allele at each of the H-2 regions (*K, I, S* and *D*).

Table 5.3 lists the haplotypes, alleles, and specificities of some inbred strains. Through genetic crosses of strains segregating from different haplotypes it is possible to find progeny representing recom-

Table 5.3 H-2 designation for several representative inbred strains

	Alleles* at loci				
Haplotype	*K*	*I*	*S*	*D*	Representative strains†
H-2^b	*b*	*b*	*b*	*b*	C57BL/10(B10), C57BL6 (B6), BALB.b
H-2^d	*d*	*d*	*d*	*d*	BALB/c, DBA/2, NZB, B10.D2
H-2^k	*k*	*k*	*k*	*k*	AKR, C3H, CBA, B10.BR, BALB.k
H-2^a‡	*k*	*k-d*	*d*	*d*	A, B10.A

* Alleles at *K* and *D* loci are defined by reactions of alloantisera (agglutinate red blood cells or lyse lymphocytes), as in the following examples (the specificities of the alloantisera and the corresponding Ags are numbered):

K^b (*b* allele at *K* locus) has private specificity 33 and public specificities 5 and 36.

D^b (*b* allele at *D* locus) has private specificity 32 and public specificity 6.

K^d (*d* allele at *D locus*) has private specificity 31 and public specificities 3 and 8.

D^d (*d* allele at *D* locus) has private specificity 4 and public specificities 3, 6, 13, 36, 41, 42, and 43.

K^k (*k* allele at *K* locus) has private specificity 23 and public specificities 3, 5, 8, 11, and 25.

D^k (*k* allele at *D* locus) has private specificity 32 and public specificities 3 and 5.

† The congenic strains: BALB.b and BALB.k, with H-2^b and H-2^k haplotypes, respectively, were derived by breeding chromosomal segments with these haplotypes into the BALB/c (H-2^d) strain. Similarly, the congenic lines B10.D2 (H-2^d), B10.BR (H-2^k), and B10.A (H-2^a) were all derived from the B10 (H-2^b) strain by a specific breeding program.

‡ The H-2^a haplotype is a recombinant: alleles at the left side of the *I* region have serological and functional properties of the H-2^k haplotype and at the right side of this region the properties of the H-2^d haplotype.

bination within the H-2 region. The frequency of recombination between two loci is used to determine to the map distance between the two markers; hence one would predict that the frequency of recombination between, for example, the *K* and *I* regions would be less than that between the *K* and *D* regions. This is what has been observed: the frequency of recombination between the *K* and *D* regions is about 0.5%, compared with 0.3% between *I* and *K*. The entire complex encompasses about 2 centimorgans (1% recombination = 1 centimorgan) of DNA. When a recombinant individual has been identified, it is possible to derive a recombinant strain and maintain the recombinant haplotype by inbreeding. Such recombinants have been essential for the genetic dissection of the *H-2* locus. In this manner the genetic positions of the various regions have been determined by analysis of multipoint recombination data.

Recombination studies have also been used to determine whether the ability of an animal to respond to two different antigens is under unigenic or multigenic control. If the immune response to antigens *x* and *y* is under multigenic control, then it should be possible to obtain recombinants that can elicit a response to antigen *x* or *y* but not to both. Conversely, if the response to both *x* and *y* is due to a single gene, it would be highly unlikely, if not impossible, that a recombinant would be obtained that could elicit a response to one antigen but not the other. Again, the response is a function of genetic map distance. Thus, through basic genetic manipulation, it is possible to study the *H-2* locus and derive congenic strains of mice differing at one genetic region of the MHC. By alloimmunization with the appropriate antigens, anti-*H-2* antibodies can then be obtained against products of the entire region or against subregions.

From such studies emerged a picture of the H-2 complex as a maze of separate loci whose interrelationships remained unclear. Thus, a reconsideration of the links between the classical genetic recombinational data and the functional characterization of various loci seems to be unavoidable.

Skin graft experiments revealed the existence of a genetic control governing this phenomenon more than 40 years ago. Many years later, it was discovered that genetically different strains varied markedly in their ability to produce antibodies (Abs) against a particular Ag; today we know that the H-2 complex is designed to control both of these activities.

The alloantigens responsible for allograft rejection map to the *K*, *I-A*, *I-E*, and *D* regions as well as those involved in cell-mediated lympholysis (CML). A third related phenomenon, the mixed lymphocyte reaction (MLR), appears to be activated by determinants controlled by loci that map predominantly in the *I-A* and *I-E* subregions. Most antigen loci controlling the immune response, the so-called *Ir* gene loci, map in the *I* region, in which are located the genes determining the high or low immune responsiveness of a strain toward a defined Ag. This region also contains the Is genes, responsible for a kind of imunosuppression. The immunosuppression in general is mediated by a subpopulation of T lymphocytes, some of which possess a surface alloantigen controlled by the *I-J* subregion (whose inclusion in the H-2 complex is now challenged; see below). The interaction of B and T helper cells also appears to be mediated by structures codified by *I*-region loci; this interaction is required for a successful immune response. A fundamental phenomenon essential for an efficient expression of the immune response is the restriction of specificity: the response of T lymphocytes that triggers the cascade of reactions leading to the success of the immune reaction is specific not only for the Ag but also for the H-2 molecules of the Ag-presenting cells. These molecules are coded in *K, D,* and *I* regions.

Finally, the genes controlling complement activity also map in the H-2 complex (*S* region). However, because the *S* region does not control any trait codified in the *K, D,* or *I* regions, nor is it controlled by any product of these regions, it can be considered evolutionarily an "alien immigrant" into the complex.

The H-2 gene products. The products of *K*- and *D*-region genes are called H-2 or class I antigens, whereas those controlled by the *I*-region genes are called I-associated or Ia or class II antigens. That the H-2 was not a single gene was discovered from the observation that H-2$^{k/d}$ heterozygotes accepted grafts from an H-2^{a} strain. It was thus postulated that H-2 consisted of two linked loci, *H-2K* and *H-2D*. The H-2^{a} strain should have (and indeed does have) at one locus the allele of the H-2^{k}, and at the other locus that of the H-2^{d} strain. The H-2^{a} graft was not rejected because it possessed the same histocompatibility antigens as the recipient animal.

K and D gene products. The K and D products can be further subdivided serologically into two classes of specificities, "public" and "private." Private specificities represent those antigenic properties of the H-2 antigens that determine their immunological identity. These specificities are uniquely determined by the *K* or *D* locus of each haplotype. Public specificities are characterized by cross-reactivity between the K and D gene products. These specificities could be attributed to either the *K* or *D* locus, or even both. Because the genes at both the *K* and *D* regions of the *H-2* locus exist as autosomal codominant alleles, it is possible to have four H-2 antigens present on the cell surface of a heterozygote, each with a private specificity. These molecules exist on the cell surface as independent species, as demonstrated by "capping" experiments,* in which each of the H-2 antigens could, with a specific antiserum, be demonstrated to cap independently of the others. It is also possible to precipitate individual H-2 antigens from solubilized proteins independently of other H-2 antigens present, thus facilitating the isolation and chemical characterization of the antigens.

The molecular complex of class I antigens (which are present on the surface of all the cells of the organism) consists of a glycosylated heavy chain embedded into the cell membrane in a close but not covalent association with β2-microglobulin, a protein codified outside the MHC. The genetic part of each molecule is exposed on the extracellular side of the membrane (Fig. 5.1) and bears the alloantigenic determinants. Only the heavy chain crosses the lipid bilayer with a short sequence of hydrophobic amino acids which apparently has an anchorage function.

The heavy chain has a molecular weight of 45,000, the light chain 11,500. The class I antigens are identified and purified by the use of specific antisera both directly (precipitation) and indirectly (inhibition of cytolysis). The primary structure of several class I antigens is known. Interestingly they enclose a region homologous to the constant region of Ig and to β2-microglobulin, but apparently they lack a defined region comparable to the hypervariable region of Ig. It was noted that the allelic products differ among themselves less than those of different loci; however, structural features related to a particular allelic series are not evident. The sequence of β2-microglobulin shows conspicuous homologies with the constant region of Ig so that it was postulated that this protein corresponds to a free-Ig domain. The recent construction of a cDNA clone for mouse β2-microglobulin should allow a more precise comparison with the Ig genes. The K and D gene products are decisive for the recognition by cytolytic T lymphocytes of cells presenting foreign an-

* Capping is defined as the migration of surface molecules in the fluid plane of the membrane to one pole of the cell. In these experiments, migration is induced by antibodies to H-2 reacting with H-2 molecules on the surface.

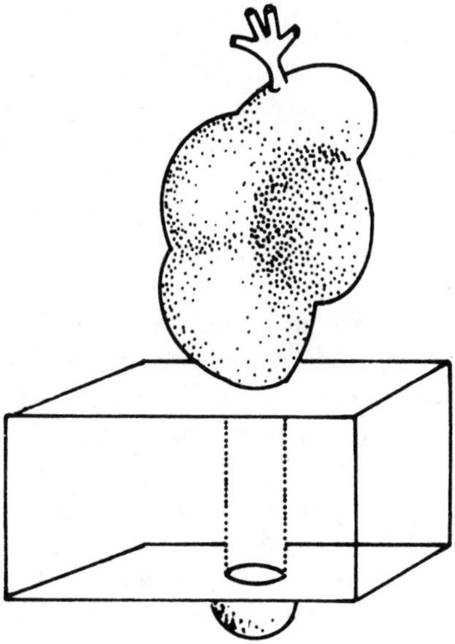

Fig. 5-1 A hypothetical model for a class I antigen

tigens on their surface. Cytotoxic T cells elicited by virus-infected cells, hapten-modified cells (Shearer et al., 1975) or minor histocompatibility antigens (Bevan, 1975) will kill only cells bearing the same foreign antigen and the same H-2 antigens of the sensitizing cells. This H-2 restriction appears to be of paramount importance for self-recognition.

I-*region gene products*. The *I* region of the H-2 complex was recognized through its control over immune responses to many Ags; the ability to produce high Ab titers to certain Ags was shown to be associated with various H-2 haplotypes. For instance, mice homozygous for $H\text{-}2^k$ are highly susceptible to Gross leukemia virus, whereas those with the $H\text{-}2^b$ allele are quite resistant. Through breeding experiments it was demonstrated that the controlling locus, called *Ir-1*, lies in the *I* region, between the *K* and *D* loci. Class II (or Ia) antigens have a much more restricted distribution than class I molecules. Ia antigens are found on B lymphocytes, macrophages, some sets of T cells, and a few other cells (dendritic cells, Langerhans cells). Their presence on macrophages was related to some important role in Ag presentation. The products of the I region were isolated from mouse spleen cell extracts by immune precipitation with alloantisera raised by cross-immunization of mice identical for *H-2D* and *H-2K* but different for *I* and *S* regions. Two glycoproteins of 33–36 and 25–30 Kilodaltons, respectively, were obtained.

However, in respect to the chains, the α and β chains of class I molecules have similar features and are both encoded in the MHC (in contrast with β2-microglobulin). Some experimental results seem to point to a closely similar molecular weight of the two chains, the difference depending on the extent of glycosylation. Up to now a structural definition has been obtained only for the products of the *I-A* and *I-E* subregions. Neither the *I-J* encoded molecules nor the products of the *I-B* subregion have been biochemically characterized. The α and β chains exist as dimers on the cell surface, but peptide map analysis of the tryptic fragments does not indicate an extensive homology between them. Data on genetic mapping of α and β chains of *I-A* and *I-E* regions indicate that $A\alpha$, $A\beta$, and $E\beta$ chains are encoded in the *I-A* region, whereas the $E\alpha$ peptide is encoded in the *I-E* subregion.

Genetic control of specific immune responses (Ir genes). The immune response is largely under the control of antigen-specific immune response (*Ir*) genes, which control the recognition of an antigen and the subsequent magnitude and nature of the immune response to it. This specificity control is exquisitely discriminatory. An animal that responds poorly or not at all to one antigen may be an excellent responder to another antigen that differs by only a single amino acid. Responses that are under *Ir* gene control are mostly T-dependent, in that T-cell participation is essential. The Ir phenomenon seems to be promulgated by the T-helper subset. Thus, the Ir effect is usually reflected in the switch from the primary (IgM) to a secondary (IgG) response. Several methods have been employed to restrict the heterogeneity of the immune response so that detailed genetic analysis of the Ir phenomenon could be exploited. These include:

1. Use of structurally and chemically defined antigens, i.e., synthetic copolymers such as (T,G)-A-L or (Phe,G)-Pro-L.
2. Use of autologous molecules (serum albumin) coupled with haptenic determinants (DNP).
3. Use of extremely low doses of antigen that promote activation of high-affinity clones of cells.

4. Use of phylogenetically related antigens having fewer "foreign" determinants.

The pioneering studies of Benacerraf and co-workers using the hapten-carrier conjugate 2,4-dinitrophenyl-poly-L-lysine (DNP-PLL), McDevitt and co-workers using synthetic copolymers, Rosenthal and co-workers with alloantisera directed against the MHC of guinea pigs, and a host of other research groups have helped to establish the following points concerning the histocompatibility-linked *Ir* genes.

1. They regulate the response to T-dependent antigens.
2. They map in the *I* region of the histocompatibility complex.
3. They function independently of idiotype on Ig V region genes.
4. Dominant inheritance of high responsiveness is common, although other modes have been found.
5. They usually determine high or low responsiveness, rather than all or none.
6. Complementation of *Ir* genes in which two different *Ir* genes are needed for a response to a single determinant have been detected.
7. Soluble factors released from T cells have been produced from 'helper' and 'suppressor' T cells. These antigen-specific factors bear Ia determinants and are functional. Some suppressor factors carry *I-J* region determinants, whereas helper factors carry *I-A* region determinants.
8. The Ir effect—and thus the *Ir* genes—is associated with the class II genes of the MHC in all species thus far studied.

Although these observations are well founded in experimental fact, concepts on mechanisms of the Ir phenomenon undergo continuing extensive revision. The postulate that the *Ir* gene coded for the T-cell antigen-specific receptor was supported by certain data, including items 1, 2, 3, and 7 above. But if *Ir* genes code for actual T-cell receptors, as many distinct *structural Ir* genes should exist as there are genetically controlled responses. Class II molecules cannot account for the specificities of all postulated *Ir* genes.

An alternative, known as the associative-recognition hypothesis, proposes that antigen must be recognized in close association with class II molecules (or Ia antigens) on the surface of the macrophage. This complex would then be the unique entity that is recognized by T cells. This viewpoint is supported by the fact that Ia antigens are distributed on such antigen presenting cells, i.e., macrophages, dendritic cells, Langerhans cells, microglial cells, and Kupffer cells. Antisera directed towards these Ia molecules can abrogate antigen recognition by T cells. Also, T cells of an F1 hybrid between poor and good responders are stimulated to antigen-specific proliferation only when the antigen is presented on F1 or responder parent macrophages. Additionally, the Ia molecules must be highly polymorphic. This heterogeneity of class II molecules would enhance the capability to associate with many important antigens in which a poor responsiveness might prove lethal. Complementation (*cis* or *trans*) produces eight distinct molecules in any given heterozygote. If each combinatorial determinant acts as a unique-antigen-binding site, then the extent of variability may be sufficient for recognition of a wide spectrum of antigens. Finally, structural changes in Ia molecules due to mutation have revealed that such change has modified the Ir control of the responses to some antigens.

Biological significance of H-2 genes and gene products. The classical interpretation of H-2, set forth when little was known about the biological significance and physiological role of *H-2* loci, led to the impression of an incomprehensible group of loci, regions, and subregions that were interrelated by obscure links. Today it is possible to present a more simple and unifying picture of this genetic complex.

It is apparent that *K* and *D* loci control substantially the same activities: a rapid allograft rejection, a strong cytotoxic reaction (CML), a weak mixed reaction (MLR), and a specifically restricted activation of cytolytic T lymphocytes. All of these activities could be ascribed reasonably to one 45-Kilodalton polypeptide. Genetic evidence on the basis of mutation studies favors this hypothesis. As to the *I*-region genes, illuminating results arose from the work on murine strains endowed with a "nonexpressor" allele at the *E* locus, an allele that prevents the appearance of E molecules on the cell surface (Juretic *et al.*, 1981); thus a positive CML is obtained only if the stimulator cell expresses the E molecule, whereas it is impossible to elicit any response toward cells lacking it. Similar results were obtained by the use of Ia-specific antibodies against the A molecules. Almost identical results came from MLR experiments using the same mutants, thus indicating that the genes and gene products involved are very much the same.

The functional link between the *I-A* and *I-E* regions was clarified with the *in vitro* T-cell proliferation assay, which consists of sensitizing an animal with a given antigen, collecting T cells several days later, and culturing them in the presence of macrophages plus the relevant antigen (T-cell proliferation is measured as [^{3}H]thymidine uptake). With a polymer of glutamic acid and alanine (GA), it was observed that the response in the test was controlled by the *I-A* subregion. The response against another antigen (GLT, a polymer of glutamic acid, lysine, and tyrosine) appeared to be controlled by two *Ir* genes, mapping in the *I-A* and *I-E* regions, respectively. These genes gave rise to a sort of complementation: namely, from crosses between nonresponder strains, high-responder recombinants were obtained. The *in vitro* T-cell proliferation assay revealed that this response is controlled by the E molecule, codified, as it is known, in *I-A* for the β subunit and in *I-E* for the α subunit. The A and E molecules can be responsible even for the activation of T helper lymphocytes. Thus the whole control of the immune response, seen in a more unitarian view, could be attributable to the A and E molecules, which are the products of the *Ir* genes.

As to the remaining *I* subregions, *I-B*, whose products are still unidentified, was reported to control the immune response against a few antigens such as LDH_B and IgG2a, but does not influence MLR, CML, or graft rejection. However, the response to LDH_B and IgG2a are explainable just on the basis of A- and E-molecule action, and the need for the *I-B* region is uncertain. The *I-J* region was long postulated to be the site for the control of immunosuppression, but the only evidence for this role was the presence of *I-J* determinants on the surface of a proportion of suppressor T lymphocytes. An interesting and recent study using cDNA clones demonstrated that the *I-A* and *I-E* genes are adjacent, separated possibly by less than 1 Kilobase and at most by 3.2 Kilobases. Thus there seems to be little, if any, space for the *I-B* and *I-J* subregions. The authors offer some explanations to reconcile their finding partly with the preceding models of the H-2 complex, but one simple possibility is that the *B* and *J* genes do not exist (Klein and Nagy, 1982).

To give support to a simpler model of the H-2 complex (Table 5.4), one must allow the possibility that various traits controlled by this chromosomal segment might not have any biological significance. A mouse in its normal life will neither undergo transplants nor receive lymphocytes from another mouse. Other traits display an unchallenged physiological value: the immune response, immune suppression, T–B cell interaction, and restriction of T-lymphocyte specificity. All of these traits can be reconciled to the basic phenom-

Table 5.4 A simple model of the H-2 complex

Locus	*K*	*A*	*A*	*E*	*E*	*D*	*L*
Molecule*	45K	28K	32K	38K	32K	45K	45K
Functional class	I	II	II	II	II	I	I

* Designated by molecular weights in thousands (e.g., 45K = 45,000).

enon of the recognition of antigen as nonself by T lymphocytes—a recognition that presupposes the interaction between antigen and H-2-controlled molecules.

H-2 receptors and polymorphism. The simultaneous recognition of antigens and H-2 molecules restricts T lymphocytes' specificity, and such restriction is probably the fundamental function of the H-2 complex. The high degree of polymorphism of class I and class II molecules (over 100 alleles each for the *K* and *D* genes and at least a comparable level of polymorphism for *A* and *E* loci) optimizes the repertoire available for "associative recognition" in a population. In such a multigenic polymorphic system a viral mutant, for instance, has a very limited probability of not finding an MHC antigen with which to interact. Thus the conservation of this "bank" of polymorphic genes within the hereditary material of the species is selectively advantageous and not the genetic burden that polymorphism implies.

The Mendelian inheritance of MHC genes suggests that they are germ-line, single-copy genes. However, tumor cells may express so-called alien histocompatibility antigens, that is, antigens whose genes are apparently not included in the genotype of the organism in which the tumor developed. These findings are compatible with the hypothesis that one individual possesses more than two alleles for each MHC locus (as stated by the classical hypothesis), but only a few of these are normally expressed (for review see Parmiani *et al.*, 1979). The "DNA-cloning army" will certainly be essential in resolving this debate.

How the molecules coded by the MHC exert their function is still an open question, but two types of models have been proposed. The dual-recognition model postulates two separate receptors on the T-cell surface, one specific for H-2 antigens, the other for non-H-2 antigens. The receptor for H-2 could either be itself an H-2 coded product or an antibody gene product. The alternative, the altered-self model, postulates a single receptor on the T-cell organism and therefore (nonself) antigenic configuration determined by the interaction of a foreign antigen with the H-2 molecules.

At the beginning of its demonstration, restriction was seen as synonymous with self-recognition. Subsequently it was demonstrated that identity at the histocompatibility level between target and effector cells is not necessary for recognition of self; thus T cells learn to recognize as self the MHC antigens exposed on the thymus cells during the processing of T-cell precursors through this organ. This is best explained by a dual-recognition hypothesis.

The receptors for the foreign Ag are perhaps encoded by V_H genes, and their diversity might arise from mechanisms similar to those that generate Ab diversity. The receptors for MHC molecules seem to be more obscure. Jerne (1971) hypothesized a T-cell receptor specific for each allelic form of the MHC antigens. The hypothesis seems to disregard the impossibility by an individual to "know" the extent of polymorphism of the MHC genes within the species. Thus other hypotheses by several authors (see references) have been presented.

General comments. The H-2 and Ia systems have been subjected to intensive research that has yielded important basic information. The H-2 complex has been genetically analyzed to the extent that it probably represents the most clearly understood segment of a chromosome from a vertebrate species. The fine-genetic analysis of the H-2 complex is second only to the work and information obtained on the *Drosophila* genome. The evolution of the MHC and its relation to the Ig genes is still not clear, although they probably diverged late in evolutionary development.

The MHC System in Humans

For a number of years it has been known that a human histocompatibility system (HLA) exists which, like the H-2 system in mice and the histocompatibility systems in other mammalian species, is extremely complex and consists of a series of many closely linked genes (Fig. 5.2). Each gene corresponds to a locus, that is, it occupies a definite position in the chromosome. Each gene consists of a DNA segment responsible for the synthesis of a polypeptide chain. The chromosomal region encompassing the HLA system is located on the short arm of human chromosome 6, very close to the loci

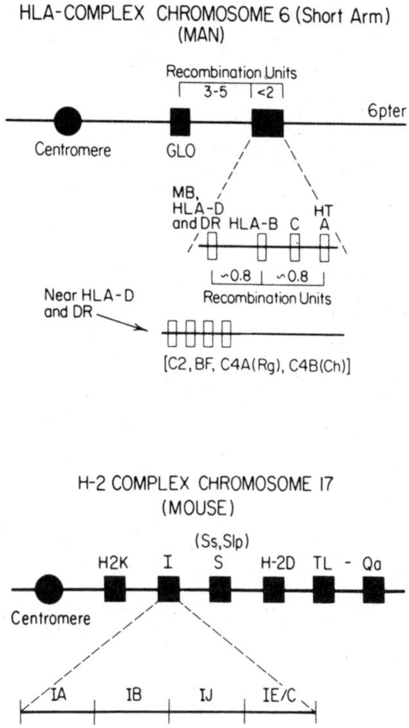

Fig. 5.2 Chromosome 6 in humans and chromosome 17 in mice for respective MHC antigens and complement proteins. (By permission, D. Raum *et al.*, Major histocompatibility markers in disease. *Clin. Immunol. Allergy* 1: 305, 1981. W. B. Saunders Co., Ltd., New York.)

coding for the malic cytoplasmic enzyme (ME-I), for urinary pepsinogen-5 (Pg5), for erythrocyte isoenzymes phosphoglucomutase-3 (PGM3), and glyoxylase (GLO). The complement system also maps on chromosome 6, at least for the C2 and C4 components. As in other mammals, this complex of genes is termed the major histocompatibility complex (MHC). As of 1972, four loci had been identified in the HLA system: *LA*, *4*, *AJ*, and *MLR* (mixed lymphocyte reaction). The antigens codified by *LA*, *4*, and *AJ* loci were identifiable by serological methods such as complement-mediated lymphocytotoxicity, whereas those codified by the *MLR* locus were identified by a cellular reaction (lymphocytes cocultured *in vitro* with allogeneic cells). The loci were then named *SD1*, *SD2*, *SD3* (SD = serologically defined), and *LD1* (LD = lymphocyte-defined).

In 1975 the WHO Committee simplified and rationalized the nomenclature, giving these loci the new designations *HLA-A, B, C,* and *D*. The HLA system also codes for DR (or Ia) antigens, a group of antigens present only on B lymphocytes, activated T lymphocytes, monocytes, macrophages, and sperm, and absent from platelets, fibroblasts, hepatocytes, and various other cell types. In the 1977 workshop, by the use of sera obtained either from multiparous women (absorbed with platelets that present HLA-A, B, C antigens but lack DR, and thus are able to remove anti-HLA-A, B, C antibodies) or from voluntary recipients of "planned immunizations" identical to the donors for HLA-A, B, C, it was demonstrated that DR antigens are controlled by genes closely linked to the *D* locus.

The name DR (meaning D-related) stems from the fact that these antigens are recognized by sera that block the MLR elicited by the corresponding D antigens. The *DR* locus-defined antigens are DR1, 2, 3, 4, 5, 6, 7, 8, 9, 10, 13; some, however, are not thoroughly characterized. Each locus of the HLA region corresponds to a series of several alleles coding for different histocompatibility antigens. At present, 90 antigens have been identified (20 at A locus, 42 at *B* locus, 8 at *C* locus, 12 at *D* locus, and 10 at *DR* locus (Table 5.5).

Because the human genome is diploid, each individual has two number 6 chromosomes and two HLA haplotypes (two sets of *A, B, C, D,* and *DR* alleles, each of which expresses an antigen). Because the HLA genes are inherited in a Mendelian fashion as codominant alleles, each individual possesses and expresses two haplotypes, one of which is transmitted to each offspring. The human nucleated cells should thus present 2 antigens per locus, giving a total of 10 antigens. The genotypes of the offspring of two unrelated parents result from four possible combinations, and each descendant shares only one haplotype with each parent (Fig. 5.3); hence the possibility that a pair of siblings will be HLA-identical is 25%.

Some alleles at A, *B, C,* and *DR* loci are linked on the same chromosome with a frequency higher than that expected on the basis of a random assortment of the individual genes, whose frequency in the population is known. Such a phenomenon was attributed to a not understood preferential association between a pair of alleles but is termed "linkage disequilibrium" (a typical example is *HLA-A1* linked to *HLA-B8*).

The histocompatibility antigens HLA-A, B, C (Fig. 5.4) are glycoproteins composed by 2 polypeptide chains α (44 Kilodaltons) and β (12 Kilodaltons). The α chain penetrates through the cell mem-

Table 5.5 Complete listing of recognized HLA specificities

HLA-A	*HLA-B*	*HLA-C*	*HLA-D*	*HLA-DR*
HLA-A1	HLA-B5	HLA-Cw1	HLA-Dw1	HLA-DR1
HLA-A2	HLA-B7	HLA-Cw2	HLA-Dw2	HLD-DR2
HLA-A3	HLA-B8	HLA-Cw3	HLA-Dw3	HLA-DR3
HL-A9	HLA-B12	HLA-Cw4	HLA-Dw4	HLA-DR4
HLA-A10	HLA-B13	HLA-Cw5	HLA-Dw5	HLA-DR5
HLA-A11	HLA-B14	HLA-Cw6	HLA-Dw6	HLA-DRw6
HLA-Aw19	HLA-B15	HLA-Cw7	HLA-Dw7	HLA-DR7
HLA-Aw23(9)	HLA-Bw16	HLA-Cw8	HLA-Dw8	HLA-DRw8
HLA-Aw24(9)	HLA-B17		HLA-Dw9	HLA-DRw9
HLA-A25(10)	HLA-B18		HLA-Dw10	HLA-DRw10
HLA-A26(10)	HLA-Bw21		HLA-Dw11	
HLA-A28	HLA-Bw22		HLA-Dw12	
HLA-A29	HLA-B27			
HLA-Aw30	HLA-Bw35			
HLA-Aw31	HLA-B37			
HLA-Aw32	HLA-Bw38(w16)			
HLA-Aw33	HLA-Bw39(w16)			
HLA-Aw34	HLA-B40			
HLA-Aw36	HLA-Bw41			
HLA-Aw43	HLA-Bw42			
	HLA-Bw44(12)			
	HLA-Bw45(12)			
	HLA-Bw46			
	HLA-Bw47			
	HLA-Bw48			
	HLA-Bw49(w21)			
	HLA-Bw50(w21)			
	HLA-Bw51(5)			
	HLA-Bw52(5)			
	HLA-Bw53			
	HLA-Bw54(w22)			
	HLA-Bw55(w22)			
	HLA-Bw56(w22)			
	HLA-Bw57(17)			
	HLA-Bw58(17)			
	HLA-Bw59			
	HLA-Bw60(40)			
	HLA-Bw61(40)			
	HLA-Bw62(15)			
	HLA-Bw63(15)			
	HLA-Bw4[a]			
	HLA-Bw6			

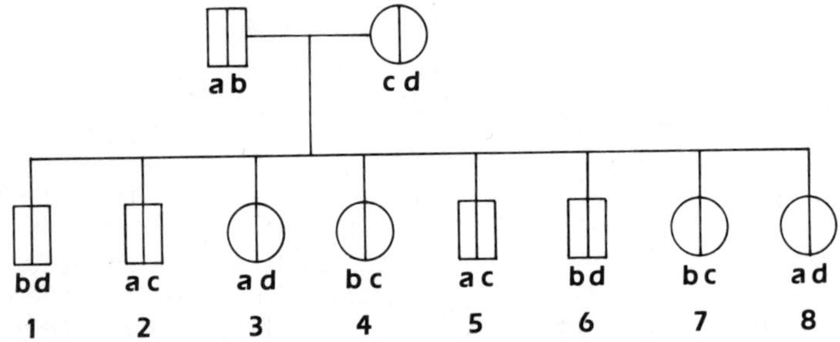

Fig. 5.3 Familial inheritance of HLA haplotypes. The father was assigned the ab and the mother the cd haplotype. The children numbered 1 and 6 are HLA-identical (haplotype bd), as in 2 and 5 (ac), 4 and 7 (bc), 3 and 8 (ad).

brane and reaches the cytoplasm with an end able to form phosphodiester "high-energy" bonds.

The β chain, termed β2-microglobulin, is controlled by a locus on chromosome 15; like the mouse β2-microglobulin (β chain of H-2 antigen), the β chain is entirely exposed on the extracellular side of the membrane and it binds the α chain via a hydrogen bond.

Fig. 5.4 Scheme of human MHC. Loci of *HLA* region that are not yet included into defined "systems" are not represented. The carbohydrate chains are indicated as —CHO.— S—S means a disulfide bridge. The molecular identification of *HLA-D* locus products is still uncertain.

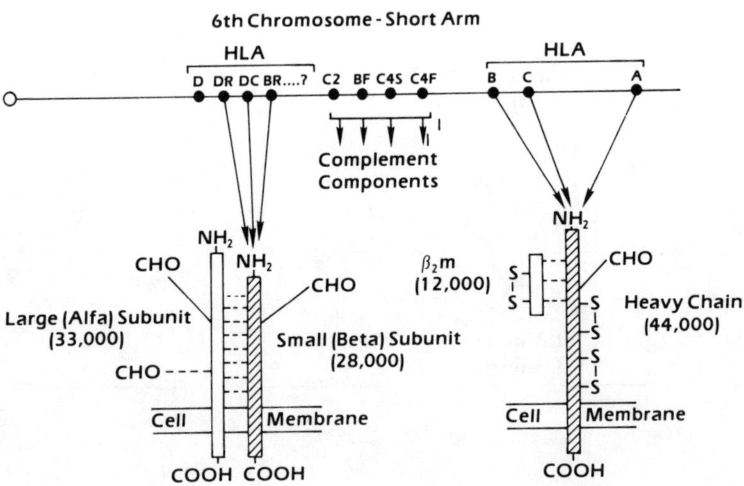

The HLA serological specificity is due to the α chain. The cells where the HLA antigens are most abundant are lymphocytes, platelets, granulocytes, and monocytes. By contrast, erythrocytes show very small amounts of HLA antigens.

Placenta, liver, intestinal, renal, and myocardial cells have different amounts of HLA antigens. The structure of HLA-D antigens, revealed by mixed lymphocyte cultures and encoded in the *D* locus, is still undetermined, but Ia antigens expressed on B lymphocytes have been very well studied. HLA-D antigens are molecules of 65 Kilodaltons consisting of two subunits (α and β) of 33 and 21 Kilodaltons linked by noncovalent bonds. Both chains reach the cytoplasm traversing the cell membrane, but only the α chain shows a phosphodiesteric high-energy bond in the intracellular part of the molecule. The smaller β subunit is certainly controlled by the *HLA* region. If the α and β chains are separated and exposed to alloantisera recognizing DR specificities, the β subunit reacts whereas the α subunit does not. Research on the human MHC has benefited from the results obtained on the murine H-2 system, including (1) the genetic control of the immune response, (2) its importance in antigen presentation and processing, (3) *Ir* genes in the interactions between T and B lymphocytes, and (4) the Ia associated antigens. From the substantial hints given by murine *Ir* genes, a comparable situation was anticipated and found in humans.

Actually the extrapolation to humans of the knowledge accumulated in mice has had a noteworthy practical importance. Precise knowledge of the mechanisms regulating the immunological defenses of the organism will make possible intervention by these same mechanisms (immune manipulation), eliciting either response (e.g., against tumors) or tolerance (e.g., abetting transplantation). In humans, an analysis was made of lymphoid cell marker molecules whose structure and distribution in various cell types could be considered analogous to the Ia molecules of the mouse. The hypothesis was that such molecules were the products of *Ir* genes and that their function was the regulation of the immune response. As in the mouse, the experimental approach utilized alloantisera. Because congenic individuals used for the immunizations are not feasible in humans, volunteers were selected for planned transfusions from donors identified for *HLA-A, B, C* locus antigens, but incompatible for new or unknown loci.

By this approach, human Ia molecules similar to those found in mice were identified, and it was demonstrated that they are controlled by the *HLA*-region genes. The existence of such a large number of alleles at the different HLA loci and, in general, the high level of "polymorphism" of the *MHC* region is a problem of great biological relevance that has not yet been satisfactorily explained. One possible partial explanation frequently appears in individuals bearing a particular HLA antigen. In 1976, the WHO promoted the Paris Conference where an International Registry was established to collect worldwide data on the association between HLA antigens and disease. The degree of association between the "marker" (i.e., the histocompatibility antigen) and the disease is assumed by comparing the frequency of an antigen in a group of patients with the frequency of the same antigen in healthy reference groups. The degree of association gives rise to the relative risk coefficient. This coefficient implies that the carrier of the antigen has a probability of becoming afflicted with a given disease 2, 4, or even 50 or more times higher than that of individuals lacking the relevant HLA antigen. Table 5.6 shows that most of the reported diseases are associated with one or another antigen coded for by *HLA-B* and *DR* loci. It is therefore probable that the genes related to the susceptibility for these diseases

Table 5.6 List of association of HLA antigens and disease

Disease	HLA antigen	Relative risk
Ankylosing spondylitis	B27	87.4
Reiter's syndrome	B27	37.0
Subacute thyroiditis	BW35	13.7
Psoriasis vulgaris	CW6	13.3
Acute anterior uveitis	B27	10.4
Idiopathic hemochromatosis	A3	8.2
Progressive autonomic failure	Aw32	28.7
Behcet's syndrome	B5	6.3
Hodgkin's disease	A1	1.4
Goodpasture's syndrome	DR2	15.9
Dermatitis herpetiformis	DR3	15.4
Addison's disease	DR3	6.3
Systemic lupus erythematosus	DR3	5.8
Rheumatoid arthritis	DR4	4.2
Multiple sclerosis	DR2	4.1
Myasthenia gravis	DR3	2.5
Coeliac disease	DR3	17.0
	DR7	4.0
Juvenile diabetes	DR3 + DR4	33.0

are located in a position bounded by the *D* and *B* loci and are in linkage disequilibrium with some alleles at these loci.

It may be of more value to determine complete HLA haplotypes rather than just a limited number of markers (such as HLA-A and HLA-B) or phenotypes alone. An excellent example of this phenomenon may be seen in congenital adrenal hyperplasia (CAH) due to 21-hydroxylase deficiency. CAH is an autosomal recessive disorder, the genes for which have been mapped in the *HLA* region on the short arm of chromosome 6 (6p). There is controversy over whether to place the gene(s) for this deficiency between *HLA-A* and *HLA-D/DR* or between *HLA-D/DR* and *GLO* (Fig. 5.5).

Several examples of evidence of genetic linkage disequilibrium (nonrandom association of HLA alleles) with 21-hydroxylase deficiency have been reported recently. The most striking example was found in the frequency of the *HLA-Bw47* allele in Sheffield, England, where it was 27.3% (compared with 0.4% in the general Eu-

Fig. 5.5 Mapping of loci to the short arm of chromosome 6 (chromosome 6p). The distance between loci are given in centimorgan (cM) units.

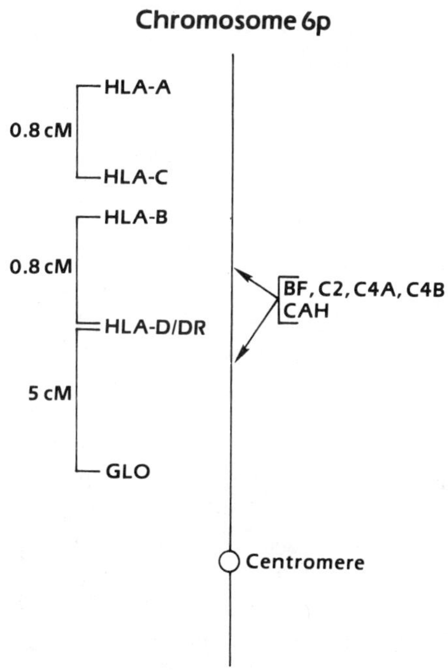

ropean population) and was also often noted to be on one particular haplotype: HLA-A3, Bw47, Cw6, FCO, 31, DR7, and GLO1. In addition, there are four complement loci, *Bf*, *C2*, *C4A*, and *C4B*, linked to the *HLA* region on chromosome 6p. Family studies indicate these to be very closely linked (no recombinants have been seen); thus the alleles present on any given chromosome can be considered a genetic unit, which has been termed a complotype. Recent studies indicate that the *HLA-Bw47* and a rare complotype *BF*F*, *C2*C*, *C4A*QO*, *C4B*31* (abbreviated FCO,31) are in genetic disequilibrium and form a part of an extended haplotype that includes the GLO allele, *GCO1*. The frequency (21.7%) of the rare *C4B* allele *C4B*31* among CAH patients was most remarkable. It was not detected in the non-disease-associated typings in the parents, and its frequency in the control population was only 0.15%. The unexpected linkage disequilibrium between *GLO* and the major histocompatibility complex alleles is likewise remarkable. The 6-centimorgan distance between *GLO* and *HLA-B* would indicate an expected recombination rate of 6% per generation. Thus some haplotypes may have some unusual biological properties that have kept them intact over many centuries. On the basis of these observations, it may be that extended haplotypes can serve as markers for certain genetic mutations and eventually allow the prediction of severity of an infant's disease from his genetic type.

The HLA-linked diseases affect only particular body systems, such as the skeleton (e.g., ankylosing spondylitis, Reiter's syndrome, rheumatoid arthritis), the neuromuscular system (e.g., multiple sclerosis, myasthenia gravis), the endocrine system (e.g., insulin-dependent juvenile diabetes, adrenocortical insufficiency), and the gastrointestinal system (e.g., coeliac disease, chronic hepatitis). Most (if not all) of these disorders are immunological in nature. In some cases the associations are so stringent as to constitute a useful diagnostic tool: for example, ankylosing spondylitis is associated with B27 antigen (relative risk coefficient of 87), and more than 90% of patients suffering from coeliac disease are positive for DR3 or DR7 or both.

It is of considerable interest that although of still undefined etiology many of these diseases may have an (auto-)immune component. In humans, it has been very difficult to demonstrate experimentally the presence of *Ir* genes. Could these diseases be true "experiments of nature" depending on the presence of specific *Ir* genes? For instance, juvenile diabetes is purportedly caused by a Coxsackie virus followed by autoimmunity to the pancreas. Is the association with

DR3 and DR4 perhaps the "sign" of the presence of an *Ir* gene that determines the lack of response or an erroneous (autoimmune) response to these viruses? Similarly, does the association of a coeliac disease with an *Ir* gene determine the immune recognition of gluten or of a product of gluten digestion? The study of the associations of DR diseases, besides its diagnostic, prognostic, and preventive value, may offer a highly suitable experimental system to clarify the mechanism of action of the Ia system in humans.

Experiments performed on purified Ia molecules isolated from lymphoid cell membranes demonstrated the presence of Ia molecules endowed with determinants coded by a non-*DR* locus; this antigen, named DC1, is present with high frequency when *DR* loci are present at the *DR1* or *DR2* or *DRw6* alleles, and constitutes another example of linkage disequilibrium. In addition, Ia molecules bearing a determinant, named BR 4x7, controlled by an allele at a non-*DR* locus, in linkage disequilibrium with *DR4* and *DR7* alleles have been demonstrated. On the basis of the above studies on the specificity of various alloantisera, a general scheme of the alloantigen specificities of Ia molecules has emerged that is clearly still incomplete (there are several blanks, i.e., undefined allelic specificities) but fits well with the experimental results. The scheme proposed two *HLA* loci, in addition to *DR*, involved in controlling Ia molecules. The hypothesis is based not only on cross-absorption experiments but also on an immunochemical analysis of the Ia molecules after pepsin digestion and microfingerprinting. These studies demonstrated that DC1 molecules (e.g., *DR2* and *DRw6*) differ from DR molecules in both subunits. This result presents a further homology with the mouse, because the A and E molecules differ between them in both subunits. As the E molecules are thought to be homologous to the human DR, the A molecules could be homologous to the human DC. One recent result is that BR 4x7 molecules (certainly non-DR molecules, on the basis of cross-absorption experiments) substantially differ from DC1 in both subunits. Therefore these molecules should be controlled by a third locus, distinct from both *DR* and *DC1*. Molecules bearing specificities allelic to one and the same locus are different (both in humans and in mice) in only one subunit, the smaller or β. Differences at both subunits constitute strong evidence for a control depending on different loci. Some recent data suggested that the products of the three loci are distributed unevenly among various cell populations displaying different functions. Recently, specific alloantigenic determinants have been identified on the products of these three loci, and it will be possible to attempt functional assays to gain more insight into this problem.

The multiple-linkage disequilibrium between the alleles of these three loci could have an important biological significance. There is a continuing debate about the explanations of linkage disequilibrium. Does the preferential association between alleles at different loci correspond to a selective advantage, or is it the consequence of phenomena such as migration and genetic drift? In fact, in the Ia system the multiple-linkage disequilibrium, besides its usefulness in the analysis of the system, could enhance understanding of some DR–disease associations. In Table 5.6, some diseases can be seen to be associated with two *DR* alleles instead of one. For instance, coeliac disease is associated with *DR3* and *DR7*. Does this mean that the susceptibility to this disease is related to the presence of both *DR3* and *DR7*, which possess a similar or even synergistic effect (excess of 3/7 heterozygotes)? Or does it mean that the Ia molecule (if any) controlling the disease is the product of an allele to a third locus associated in linkage disequilibrium with both *DR3* and *DR7*? The identification of this still hypothetical molecule (e.g., by the analysis of antisera giving cross-reactions 3–7) might answer this question. The recently discovered association between *DC3* in linkage disequilibrium with *DR3* or *DR7* and coeliac disease sheds new light on the role of the various loci and antigens of the MHC.

CELL INTERACTIONS IN THE EVENTS INVOLVED IN THE SPECIFIC IMMUNE RESPONSES

At least three types of cells are involved in the immune response: B cells, accessory cells or macrophages, and T cells. The B cells are the precursors of antibody-forming cells; the T cells do not synthesize immunoglobulin but are necessary for stimulating the B cells to differentiate to produce antibody; and the macrophages are necessary for an immune response to occur in collaboration with both T and B cells. However, some antigens are thymus-independent (e.g., polyvinylpyrrolidone), dextran, and *Escherichia coli* lipopolysaccharide, in that they will interact directly with B cells. Antigens such as erythrocytes, serum proteins, and hapten–carrier complexes are strongly thymus-dependent. In this case it is generally thought that helper T cells recognize the carrier portion, whereas B cells react to the hapten.

Direct detection of antibody production by single cells can be accomplished by the hemolytic plaque technique. For example, after immunizing a mouse with sheep red blood cells (SRBC), various organs (such as the spleen or lymph nodes) containing antibody-forming cells are removed and prepared as a single cell suspension. This suspension is then mixed with soft agar that contains the antigen (in this case, SRBC) and is placed on glass slides. After the agar hardens, the slides are incubated at 37°C and complement is layered over the slides; small clear areas or plaques are formed wherever there are cells that have produced hemolytic anti-SRBC antibody.

From the number of plaques one can estimate the actual number of antibody-forming cells in a particular organ. The assignment of lymphoid cells from various organs to distinct roles in antibody formation is made with the adoptive-transfer technique. In this procedure, mice are lethally irradiated, thereby depleting their own lymphoid system and rendering them unable to mount an immune response. Then donor cells from genetically identical mice, either from individual organs or as mixtures, are injected into the irradiated host. Simultaneously with cell injection one also introduces the antigen (SRBC); after an appropriate time, antibody-forming cells can be enumerated by the hemolytic plaque assay. That the cells that produce antibody are in fact donor cells and not surviving host cells is proved through the use of specific chromosome markers and alloantibodies.

With these methods it has been demonstrated clearly that a true synergism exists between T and B cells in the response of mice to SRBC (see Fig. 5.6). Thus, after adoptive transfer with T and B cells, the antibody response is greater than the sum of the responses for T or B cells alone. In fact, neither normal nor immune T or B cells transferred alone make any significant amount of antibody. However, normal spleen cells contain all of the crucial elements needed to form antibody, and presumably contain an immunocompetent mixture of T and B cells. The synergistic response of T and B cells can be related logarithmically to the number of thymus and marrow cells contained in the cell mixture transferred, when one of the cell populations is in excess and the other is limiting.

In further experiments it was shown that contact between antigen and T cells was required for antibody formation. Furthermore, the antigen–T cell complex that cooperated with B cells to produce antibody was radiosensitive. The latter finding supports the notion that a viable T cell, rather than a processed antigen associated with

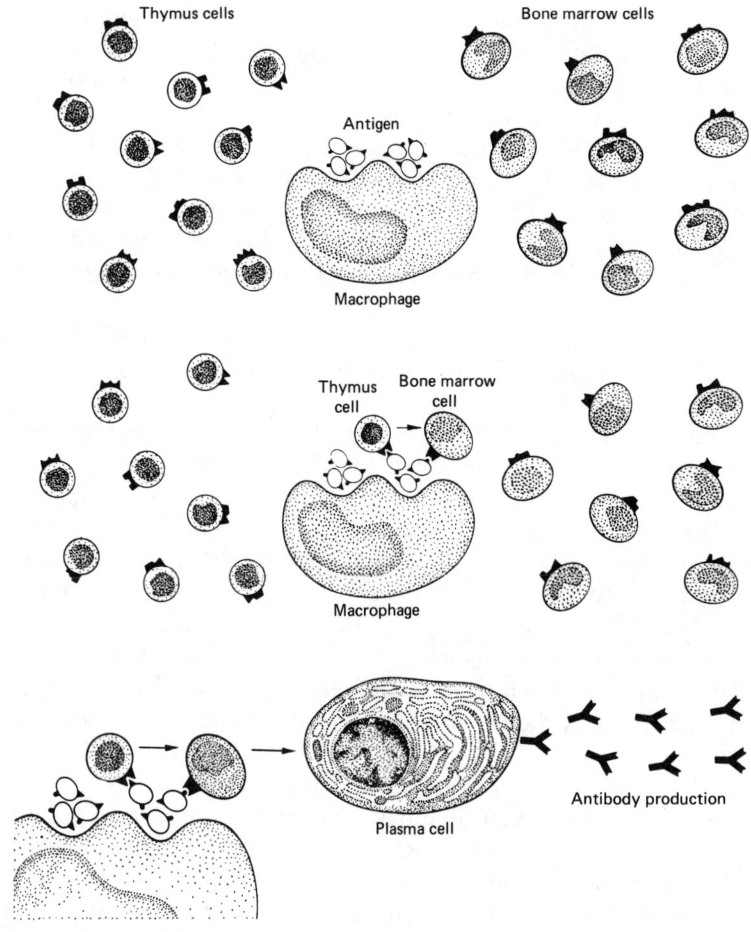

Fig. 5.6 Thymus-derived cell and bone marrow cell interact synergistically in the development of an antibody response. As schematized, (1) the macrophages first fix antigen on their surface membranes. Specifically committed cells of both lymphoid populations then attach to different determinants on the aggregated antigen. As a result of this interaction with antigen, the bone marrow and thymic cells are brought into close proximity for a sufficient period of time to permit passage of information from the thymic to the bone marrow cell (2). This interaction provides a signal to the bone marrow cell, probably in the form of a change in the surface membrane configuration. In response to this signal, the synthetic mechanisms of the marrow cell are activated and the cell differentiates into plasma cells that elaborate the specific antibodies (3). (By permission, W. O. Weigle, Immunologic unresponsiveness. *Hosp. Practice* 6: 121, 1971. Hospital Practice Co., New York)

a dead cell, must function in this interaction. Later studies showed that this T-cell "helper" effect becomes radioresistant after priming with antigen. The classic investigations of Miller and Mitchell on cell–cell interaction in the immune response have shed much light on the roles of T and B cells in antibody formation. Neonatal thymectomy in the mouse virtually abolishes the antibody response to SRBC in later life. Injected thymus or thoracic duct lymphocytes, and to a lesser extent allogeneic T cells, can restore this response even if they are derived from semiallogeneic mice. However, B cells, irradiated normal T cells, or thymic extracts were ineffective in this regard.

In further experiments the synergistic effect of mixed populations of T and B cells on the response of mice to SRBC was confirmed. Although thoracic duct lymphocytes have been considered to be primarily T cells, they can produce responses equal to those seen with thymus and marrow cell mixtures. By use of selective ablation of B cells with alloantibodies, it could be shown that the antibody-forming cells in thoracic duct lymph were B- and not T-derived. These experiments clearly show that at least two cell types, one thymus-dependent and the other bone marrow-dependent, must cooperate to restore immune competence for SRBC in lethally irradiated, neonatally thymectomized mice.*

It has also been possible to analyze cell cooperation in the response to SRBC *in vitro* using the hemolytic plaque assay on spleen cells that are cultured *in vitro* for brief periods of time with SRBC. Using this technique, Mosier found that at least two cell types, and possibly three, are required in spleen cultures for an adequate antibody response to SRBC. Pierce extended these observations and showed that one cell type is a macrophage or accessory cell. Interaction of macrophages with lymphoid cells is an obligatory requirement for the formation of cells producing anti-SRBC antibody derived from nonimmunized mice. However, if the animals are "primed" with SRBC before establishing spleen cell cultures, this requirement is either drastically reduced or abolished.

The pretreatment of the spleen cell suspension employed in the *in vitro* assay for SRBC antibody with a cytotoxic antiserum directed against thymus cells (i.e., against Thy-1 antigen—formerly θ antigen) greatly reduces the plaque-forming cells themselves, but appears to

* For certain other antigens, particularly polymerized substances, the requirement for T cells is far less stringent.

affect the T cell required for initial antigenic contact. The response of spleen cells pretreated with anti-Thy-1 can be completely restored by introducing normal thymic lymphocytes. These experiments provide additional evidence for two cell populations interacting in an *in vitro* system for antibody production.

Some additional comments on the role of macrophages in the induction of humoral immune responses seem warranted. Antigens may be phagocytized and "processed" by the cells before any contact with the T or B cell. Evidence that extracts of macrophages containing RNA can transfer immunological memory to lymphoid cells has been presented but remains controversial. It is likely that the role of the macrophage is to present antigen in such a way that T- and B-cell cooperation is facilitated. This process could occur on the surface of the macrophage, thereby allowing the receptors on T and B cells to attach simultaneously to the antigen. Currently it is felt that the role of the macrophage in the induction of the immune response does not provide immunological specificity; this cell type has been reported to produce factors capable of stimulating both B and T cells. The nature of interaction between macrophages and T cells was investigated by Feldmann and co-workers using flasks with double chambers separated by cell-impermeable membranes; they found that there was effective interaction, indicating that contact between T cells and macrophages is *not* required for helper-cell induction. However, similar experiments indicate that when T and B cells are separated, an effective response can be induced so long as the B-cell compartment contains macrophages. Recent investigations of the roles of CI and other *I*-region gene products were discussed earlier in this chapter.

The most recent study of mouse and human cells involved in immune cell–cell cooperation suggests this model: T and B cells possess specific receptors for antigen and cooperate in the elaboration of an antibody-forming response to thymus-dependent antigens. This collaboration could occur by contact and interaction of T and B cells in which the T cell focuses and presents antigen to B cells of the appropriate specificity; or by a mechanism in which the T cells release a product (specific or nonspecific for the antigen) that can stimulate B cells. A strong genetic requirement for compatibility in histocompatibility antigens has been reported for cooperative interaction involving direct contact of activated T cells and specific B cells.

Antigen in association with macrophages facilitates cooperation between two subsets of T cells (suppressor and helper). Interaction among T "helper" cells, macrophages, B cells, and antigen results in B-cell differentiation into antibody-secreting plasma cells.

Antigen Receptors on Lymphoid Cells

Interaction between lymphocyte and antigen implies a receptor or recognition site for antigen on the cell surface. Such receptors could be either specific for the antigen in question or totally nonspecific, attaching with equal avidity to any antigen. The most likely candidate for a specific receptor would be an Ig or antibody-like molecule with antigen specificity and binding characteristics similar to antibody Fab sites. It has been suggested that direct contact between antigen and B-cell receptor immunoglobulin gives a signal to the cell that results in tolerance. In contrast, if the antigen were complexed with T-cell "immunoglobulin" (associative antibody), the B cell would receive a second signal that would cause it to proliferate and to form antibodies. In terms of genetic economy, it seems extremely redundant to have a complete set of cell-bound receptors for antigen that would be chemically or structurally widely divergent from antibody. However, to date most demonstrations of immunoglobulin-like determinants on T lymphoid cells have been largely indirect in that the presence of such receptors was inferred by blocking or stimulating some cell function with antisera to various immunoglobulins or immunoglobulin fragments. The significance of such demonstrations has been questioned. In the mouse, pretreatment of spleen cells with antiserum to the predominant L chain in mouse serum antibody renders them incapable of producing antibody to SRBC *in vitro*. This effect is produced by action of antiserum on B cells.

Binding of ^{125}I-radiolabeled antigens to lymphocytes can be readily measured by autoradiography. Using this method, about 1/5000 unimmunized mouse spleen cells, predominantly small lymphocytes, are labeled with a variety of antigens. The binding of radiolabeled antigen to these cells can be inhibited by antiserum to mouse M or K chains. The function of these radioactive antigen-binding lymphocytes has been assessed by radiolysis *in vitro* followed by injection of the cells into heavily irradiated syngeneic mice. When challenged with unlabeled antigen 24 hours later, no antibody production resulted. It is of interest that thymocytes did not bind ^{125}I-labeled flagellin, thus suggesting that the splenic B-cell population preferentially binds this antigen.

Further evidence that receptors for antigen are present on B cells stems from the observation that neonatally thymectomized mice have normal numbers of antigen-binding cells in peripheral organs. In addition, anti-Thy-1 serum does not bind to the same cells that bind antigens. If adult or neonatally thymectomized mice are injected with T cells from mice bearing the T_6T_6 chromosome marker, antigen-binding cells in the spleen do not contain the marker chromosome, which implies that B cells alone bind antigen. In this same series of experiments, neither anti-H nor anti-L chain sera were capable of blocking the binding of antigens to lymphocytes. Although much of the data support an antigen-binding role for the B cell, it is assumed that T cells also contain receptors; however, the nature of the T-cell receptors (Ig or non-Ig) remains controversial.

Controversy still exists over whether the specific T-cell receptor for antigen bears a close relationship to molecules coded for (at least in some respect) by the Ig system (e.g., T and B cells share idiotype and thus V_H genes). Others argue for the coding of T cell receptors by a gene set entirely different from Ig genes.

The following is usually presented as evidence that T and B cells share Ig V_H genes:

1. Anti-idiotypic sera can induce a variety of T-cell functions and this induction appears to be linked to *Igh-1*.
2. Antigen-induced T cells react with anti-idiotypic antibodies directed to serum antibodies.
3. Antibodies obtained after immunization with activated T cells bind to antibody molecules having the same specificities as the T cells used for immunization.
4. There is an allotype-restricted induction of T-cell function by anti-idiotype antibodies.
5. *I-J* products and idiotypes have been reported to coexist on the same polypeptide chains derived from suppressor T cells and hybridomas.

In the B cell, the V_H gene is in reality a mosaic of the three rearranged gene segments V_H, *D*, and J_H. The availability of DNA probes for these Ig B-cell segments has led to the scrutiny of T cells for expression of such Ig genes. If RNA containing any V_H gene segment synthesized by T cells were present, these probes should uncover them, since less than one molecule per cell complementary to the DNA would be detected.

However, when T lymphocytes, T hybridomas, peripheral T cells, antigen-specific continuous T-cell lines, and T-cell hybrids with serologically detectable idiotypic determinants were screened by such probes, no gene rearrangement or transcription of any V_H gene could be found. Thus, it appears to date that T cells do not synthesize RNA containing any V_H gene segments, which creates considerable doubt that they use the classical V_H gene segments to encode antigen receptors. This leads to the hypothesis that T cells may have their own unique but evolutionarily related variable region genes.

Recently, immunoglobulins have been isolated from the surface of B cells after surface radioiodination. In most cases, this immunoglobulin is a monomeric form of IgM (i.e., $\mu_2\kappa_2$), but IgD has also been detected on human and mouse B cells. The interrelationship of these molecules as signal receptors is unknown although ontogenetic studies suggest different functions.

DUAL NATURE OF THE IMMUNE SYSTEM

In assessing the role and mechanisms of genetic control of the immune response, experimental approaches in animals have met with considerable success. However, the study of naturally occurring immunodeficiency states, best characterized in humans, has also been very fruitful. The basic concepts that underlie the immunodeficiency approach rest on the partially proven hypothesis of a two-component immunological system. This concept arises from the undisputed observation in chickens that the development of humoral immunity depends on a hindgut organ, the bursa of Fabricius, whereas cellular immunity depends on the thymus. There have been attempts to extrapolate this duality in the immune system to mammalian species as well. It is clear in mammals that functionally distinct cellular and humoral immune systems can be identified. However, neither a morphological counterpart of the avian bursa nor an exclusively cellular immune role for the thymus has been unequivocally demonstrated. For the purpose of this discussion, it is useful to divide immunedeficiency states into humoral, cellular, and combined. It should be recognized that currently such a division is arbitrary and certainly oversimplified. Indeed, present concepts stress the importance of interaction of the two systems in most immune reactions.

In an earlier section, we introduced the concept of T and B cells and presented evidence for their interaction in antibody synthesis. One can also consider the functions of these two cell types to be at least partially restricted to humoral (B-cell) and cell-mediated

(T-cell) immune reactions. Stem cells that originate in the bone marrow give rise to these two distinct lymphocyte populations that mature under different influences.

T lymphocytes are found in great numbers in the recirculating pool of blood and thoracic duct channels. These cells can be stimulated with nonspecific plant mitogens such as phytohemagglutinin (PHA). In mice they carry a specific antigenic marker, previously termed "theta antigen," now called Thy-1. Following contact with antigen (preprocessed by the macrophages) specifically sensitized T cells proliferate. Some then go on to develop into long-lived memory cells. Others serve a variety of effector functions, many of which are mediated by the synthesis and release of soluble factors. These substances encompass a variety of activities whose chemical natures are as yet ill defined. A partial list of these factors is presented in Table 5.7.

The demonstration of the myriad of activities listed in Table 5.7

Table 5.7 Partial list of mediators of cellular immunity elaborated by activated lymphocytes after addition of antigen

1. Skin permeability factor
2. Chemotactic factors for macrophages
3. Macrophage migration inhibitory factor (MIF)
4. Macrophage-activating factor (same as MIF?)
5. Chemotactic factors for other leukocytes (neutrophils, basophils, lymphocytes)
6. Granulocytic migration inhibitory factors (LIF)
7. Growth inhibitory factors (clonal inhibitory factor, proliferation inhibitory factor)
8. Lymphocytotoxin (toxic for all cells other than lymphocytes)
9. Eosinophil chemotactic factor (ECF)—Czarnetzki *et al.* (1976)
10. Eosinophil stimulation promoter (ESP)—Greene and Colley (1976)
11. Thymocyte-stimulating factor (TSF)—Chen and DiSabato
12. Xenogeneic reconstitution factor (XRF)—Farrar and Fuller-Bonar (1976)
13. Genetically related macrophage factor (GRF)—Erb and Feldman (1975)
14. Nonspecific macrophage factor (NMF)—Erb and Feldman (1975)
15. Histamine-releasing factor
16. T-Cell growth factor (Interleukin 2)
17. B-Cell growth factor
18. Antigen-specific suppressive T-cell factor—Tada *et al.* (1975)
19. Osteoclast-activating factor
20. Collagen-synthesizing factor
21. Interferon
22. Mitogenic factor(s) for lymphocytes

has been restricted primarily to *in vitro* systems. Whether these biologically effective substances are each chemically homogeneous, and whether they play a role in cell-mediated immune reactions *in vivo* remains undetermined. Certainly the heterogeneous activities released by sensitized small lymphocytes reveal the immense complexity of the product of the B cell, antibody.

Defects in T-cell systems can be reflected in isolated cellular immunodeficiency or a faulty antibody response in those instances where T- and B-cell interaction is required (see Chapter 7).

RELATIONSHIP OF GENETICALLY DETERMINED IMMUNE DEFICIENCY, AUTOIMMUNITY, AND LYMPHOID NEOPLASIA

Paradoxically, humans with immune-deficiency syndromes, particularly hypogammaglobulinemia, are prone to develop autoimmune diseases. In such pathological states small amounts of antibody directed against the individual's own tissues may develop—for example, against red cells, producing hemolytic anemia; or against gastric cells, producing pernicious anemia. In addition, family studies of first-degree relatives of many of the patients have shown an inordinately high incidence of similar autoantibodies when compared with normal age-matched control subjects. As previously mentioned, biochemical studies of peripheral lymphocyte responses to various mitogens (PHA, concanavalin A, and pokeweed mitogen) have clearly demonstrated an autosomal recessive form of inheritance in at least one such disease, adult "acquired" hypogammaglobulinemia. In addition, in various human immune deficiency states, particularly those with T-cell defects, there is at least a 10-fold increase in the incidence of cancer and especially lymphoid malignancy.

Following Burnet's clonal selection theory, autoimmunity results from the proliferation of so-called forbidden lymphoid clones. Other authors have made the tantalizing suggestion that autoantibody-producing cells are always present and produce small amounts of antibody that remove aged or damaged tissues. In normal individuals, however, these autoantibody-forming cells are kept in check through the surveillance mechanism of suppressor T cells. When T-cell function fails, disease from either autoantibodies or perhaps cell-mediated autoimmune reactions results. In addition, continual proliferation of unchecked autoantibody-producing clones gives rise to increased rates of mutation with chromosome aberrations. Certain mutants can be antigenically negative with respect to the host, thus giving rise to a breakdown in removal of these cells, because they are tolerated as self-constituents. Such a mechanism could result in unrestricted or malignant growth in individuals without overt T-cell deficiency.

The foregoing formulation is supported by a number of lines of evidence in humans as well as in animals.

1. Normal individuals have low concentrations of autoantibodies in their serum; their frequency increases with age.
2. Neonatally thymectomized mice of strains other than NZB and BALB/c develop kidney disease indistinguishable from the spontaneously occurring disease regularly observed in the genetically predisposed strains (NZB).
3. NZB mice that do not succumb to autoimmune hemolytic anemia have a high incidence of malignancy.
4. Humans treated with immunosuppressive agents that selectively ablate T-cell responses have an increased incidence of malignancy.

It is important in evaluating these hypotheses to differentiate between true "autoantibody" and that directed against viral or bacterial antigens. Many so-called autoimmune diseases such as systemic lupus erythematosus (SLE) in humans and the NZB-NZW syndrome in mice may, in fact, be caused by immune complexes formed with viral antigens. Recent evidence suggests an increased incidence of certain HLA antigens in patients with SLE, thereby implicating genetic mechanisms.

SUMMARY

Genetic studies of the cellular basis of the immune response are still in the early stages compared with studies of immunoglobulin structure. Plasma cells deriving primarily from the bone marrow compartment in mammals are responsible for the synthesis of antibody. In addition to humoral immunity, delayed hypersensitivity or cell-mediated immune reactions form a second system critical for resistance to certain infections and the rejection of foreign tissues and tumors. Cell interaction among T cells, B cells, and macrophages is essential in the formation of antibody to at least some antigens. The interaction of antigen with immunocompetent cells occurs on the cell surface and is mediated by cell receptors that probably have an immunoglobulin-like structure. Many aspects of cellular immune reactions must now be interpreted in light of evidence showing mediation *in vitro* by soluble effector substances with a wide variety of biological functions. It is not unlikely that HLA antigens might be receptors and important in cellular communications in other cell types (e.g., neurological cells) that utilize them in more subtle and exquisite strategies of communication.

The biological role of HLA lies in part in regulation of the immune response and resistance to infectious disease. Our information on the genetics of the HLA-A and B antigens is nearly complete for Caucasians, and it seems highly probable that in a few years our information for HLA-DR, DC, and BR may equal that. Enough is already known to conclude that the polymorphism of the HLA system is extensive and that it shows striking homology with MHCs of other mammalian species. It is assumed that this polymorphism is the result of a selection pressure. Little is known about the frequency of mutation of the genes in HLA. The extreme polymorphism and consequent heterozygosity of HLA is still not understood, principally because it appears that there are no differences between populations with less heterozygosity for HLA and resistance to local infectious diseases, but they are susceptible to imported infectious agents (such as measles and influenza). Another important point is that in spite of our information on the genetics of HLA in Caucasians, very little can be said about the distribution of these determinants on different tissues. This is essential if we are to understand the biological role of HLA in disease.

RECOMMENDED READINGS

Books

Alper, C. A. (1981), Complement and the MHC. In M. E. Dorf (Ed.), *The Role of the Major Histocompatibility Complex in Immunobiology*. New York: Garland STPM Press, pp. 173–220.

Benacerraf, B. (Ed.) (1975), *Immunogenetics and Immunodeficiency*. Baltimore: University Park Press.

Bodmer, W. F. (Ed.) (1978), *Histocompatibility Testing*. Copenhagen: Munksgaard, pp. 35–82.

Ceppellini, R. (1971), Old and new facts and speculations about transplantation antigens of man. In B. Amos (Ed.), *First International Congress of Immunology (Progress in Immunology)*. New York: Academic Press, p. 973.

Cooper, M. D. and Warner, N. L. (Eds.) (1974), *Contemporary Topics in Immunobiology* (Vol. 3). New York: Plenum Press.

Davis, B. D., Dulbecco, R., Eisen, H. N., and Ginsberg, H. S. (1980), *Microbiology* (3rd ed.). New York: Harper & Row.

Dorf, M. E. (1981), Genetic control of the immune responsiveness. In M. E. Dorf (Ed.), *The Role of the Major Histocompatibility Complex in Immunobiology*. New York: Garland STPM Press, pp. 221–254.

Ferrone, S. and David, C. S. (Eds.) (1982), *Ia Antigens*. Boca Raton: CRC Press.

Ferrone, S. and Solheim, B. G. (Eds.) (1982), *HLA Typing: Methodology and Clinical Aspects*. Boca Raton: CRC Press.

Forgereat, M. and Dausset, J. (Eds.) (1981), *Progress in Immunology* (Vol. 4). New York: Academic Press, p. 239.

Gotze, D. (Ed.) (1977), *The Major Histocompatibility System in Man and Animals*. Berlin: Springer Verlag.

Katz, D. H. (Ed.) (1982), *Monoclonal Antibodies and T cell products*. Boca Raton: CRC Press.

Katz, D. H. and Benacerraf, B. (Eds.) (1974), *Immunological Tolerance*. New York: Academic Press.

Katz, D. M. (1977), *Lymphocyte Differentiation, Recognition and Regulation*. New York: Academic Press.

Klein, J. (1982), Immunology, *The Science of Self-Nonself Discrimination* (1st ed.), New York: Wiley, pp. 156–255.

Klein, J. (1975), *Biology of the Mouse Histocompatibility Complex*. Berlin: Springer Verlag.

Loor, F. and Roelants, G. E. (Eds.) (1977), *B and T cells in Immune Recognition*. Chichester: Wiley.

Parham, P. (1981), *Receptors and Recognition*. London: Chapman & Hall.

Pierce, C. W., Cullen, S. E., Kapp, J. A., Schwartz, B. D., and Shreffler, D. C. (Eds.) (1983), *Ir Genes: Past, Present, and Future*. Clifton, N.J.: Humana Press.

Snell, G. D., Dausset, J., and Nathenson, S. (Eds.) (1976), *Histocompatibility*. New York: Academic Press, pp. 181–274.

Tanigaki, N., Koyama, K., and Pressman, D. (1978), *Proteins of the Biological Fluids*. New York: Pergamon Press, pp. 641–644.

Terasaky, P. L. (Ed.) (1980), *Histocompatibility Testing*. Los Angeles: UCLA Tissue Typing Laboratory, pp. 18–20.

Van Rood, J. J., De Vries, R. P., and Brodley, B. A. (1981), Genetics and biology of the HLA system. In M. E. Dorf (Ed.), *The Role of the Major Histocompatibility Complex in Immunobiology*. New York: Garland STPM Press, pp. 59–114.

Reviews

Benacerraf, B. and Dorf, M. E. (1976), Genetic control of specific immune responses and immune suppression by I-region genes. *Cold Spring Harbor Symp. Quant. Biol.* 41, 465.

Banacerraf, B. and McDevitt, H. D. (1972), Histocompatibility linked immune response genes. *Science* 175, 273.

Benacerraf, B., Kapp, J. A., Debre, P., Pierce, C. W., and de la Croix, F. (1975), The stimulation of specific suppressor T cells in genetic nonresponder mice by linear random copolymers of L-amino acids. *Transplant. Rev.* 26, 21.

Bodmer, W. F. (1972), Evolutionary significance of the HLA-A system. *Nature* 237, 139.

Bodmer, W. F. (1973), A new genetic model for allelism at histocompatibility and other complex loci: polymorphism for control of gene expression. *Transplant. Proc.* 5, 1471.

Bodmer, W. F. and Bodmer, J. G. (1978), Evolution and function of the HLA system. *Br. Med. Bull.* 34, 309.

Cohn, M. and Epstein, R. (1978), T cell inhibition of humoral responsiveness. II. Theory on the role of restrictive recognition in immune regulation. *Cell. Immunol.* 39, 125.

Dausset, J., Degos, L., and Hors, T. (1974), The association of the HL-A antigens with diseases. *Clin. Immunol. Immunopathol.* 3, 127.

David, C. S. (1976), Serologic and genetic aspects of murine Ia antigens. *Transplant. Rev.* 30, 299.

Doherty, P. C., Gotze, D, Truichieri, G., and Rolf, M. (1976), Models for recognition of virally modified cells by immune thymus-derived lymphocytes. *Immunogenetics* 3, 517.

Esposito, F. and Nobili, R. (1982), Preconjugant interaction in ciliates: a cell recognition problem. *Dev. Comp. Immunol.* 6, 395.

Festenstein, H. and Schmidt, W. (1981), Variation in MHC antigenic profiles of tumor cells and its biological effects. *Immunol. Rev.* 60, 85.

Janeway, C. A., Cone, R. E., and Rosenstein, R. W. (1982), T cell receptors. *Immunol. Today* 3, 83.

Jerne, N. K. (1971), The somatic generation of immune recognition. *Eur. J. Immunol.* 1, 1.

Katz, D. H., (1982), Genetic control of cell-cell interaction. *Pharm. Rev.* 34, 51.

Klein, J. (1978), H-2 mutations: their genetics and effects on immune functions. *Adv. Immunol.* 26, 44.

Klein, J. (1979), The major histocompatibility complex of the mouse. *Science* 203, 516.

Klein, J., Juretic, A., Baxevanis, G. N., and Nagy, Z. A. (1981), The traditional and a new version of the mouse H-2 complex. *Nature* 291, 455.

Klein, J. and Nagy, Z. A. (1982), Trouble in the J-land. *Nature* 300, 12.

McDevitt, H. O. (1976), The evolution of genes in the major histocompatibility complex. *Fed. Proc.* 35, 2168.

Moller, G. (Ed.) (1978), Role of macrophage in the immune response. *Immunol. Rev.* 40, 1.

Mozes, E. (1975), Expression of immune response (Ir) genes in T and B cells. *Immunogenetics* 2, 397.

Nagy, Z. A., Baxevanis, C. N., Ishii, N., and Klein, J. (1981), Ia antigens as restriction molecules in Ir-gene controlled T-cell proliferation. *Immunol. Rev.* 60, 59.

Nairm, R., Yamaga, K., and Nathenson, S. G. (1980), Biochemistry of the gene products from murine MHC mutants. *Ann. Rev. Genet.* 14, 241.

Nathenson, S. G., Vehara, H., Ewenstein, B. M., Kindt, T. J., and Coligan, J. E.

(1981), Primary structural analysis of the transplantation antigens of the murine H-2 major histocompatibility complex. *Ann. Rev. Biochem.* 50, 1025.

Paul, W. E. (1976), Genetic control of specific interactions of immunocompetent cells. Speculations on mechanism and general biologic significance. *Fed. Proc.* 35, 2044.

Ploegh, H. L., Orr, H. T., and Strominger, J. L. (1981), Major histocompatibility antigens: the human (HLA-A, -B, -C) and murine (H-2K, H-2D) class I antigens. *Cell* 24, 287.

Parmiani, G., Carbone, G., Favernizzi, G., Pierotti, M. A., Sensi, M. L., Rogers, M. J., and Appella, E. (1979), Alien histocompatibility antigens on tumor cells. *Immunogenetics* 9, 1.

Schwartz, B. D. and Cullen, S. E. (1978), Chemical characteristics of Ia antigens. *Springer Sem. Immunopathol.* 1, 85.

Schwartz, B. D., Paul, W. E., and Shevach, E. M. (1976), Guinea-pig Ia antigens: functional significance and chemical characterization. *Transplant. Rev.* 30, 174.

Shearer, G. M., Rhem, I. G., and Garbarino, C. A. (1975), Cell-mediated lympholysis of trinitrophenyl-modified autologous lymphocytes effector cell specificity to modified cell surface components. *J. Exp. Med.* 141, 1348.

Silver, L. M. (1981), Genetic organization of the mouse t complex. *Cell* 27, 239.

Simpson, E. (1982), The role of H-Y as a minor transplantation antigen. *Immunol. Today* 3, 97.

Snell, G. D. (1980), The major histocompatibility complex: its evolution and involvement in cellular immunology. In *The Harvey Lectures*, Series 74. New York: Academic Press.

Strominger, J. L., Engelhard, V. H., Fuks, A., Guild, B. C., Hy fil, F., Kaufman, J. F., Korman, A. J., Kostyk, T. G., Krangel, M. S., Lancet, D., Lopez de Castro, J. A., Mann, D. L., Orr, H. T., Parham, P. R., Parner, K. C., Ploegh, H. L., Pober, J. S., Robb, R. J., and Shackelford, D. A. (1981), Biochemical analysis of products of the MHC in the role of the major histocompatibility complex in immunobiology. In M. E. Dorf (Ed.), *Role of the Major Histocompatibility Complex in Immunobiology*. New York: Garland STPM Press, p. 115.

Uhr, J. W., Capra, J. D., Vitetta, E. S., and Cook, R. G. (1979), Organization of the immune response genes. *Science* 206, 292.

Williamson, A. R. (1980), Three-receptor, clonal expansion model for selection in the thymus. *Nature* 238, 527.

Zinkernagel, R. M. (1979), Associations between major histocompatibility antigens and susceptibility to disease. *Am. Rev. Microbiol.* 33, 201.

Zinkernagel, R. M., Callahan, G. N., Klein, J., and Demert, G. (1978), Cytotoxic T cells learn specificity for self H-2 during differentiation in the thymus. *Nature* 271, 251.

Zinkernagel, R. M. and Doherty, P. C. (1979), MHC-restricted cytotoxic T cells: studies on the biological role of polymorphic major transplantation antigens determining T-cell restriction, specificity, function and responsiveness. *Adv. Immunol.* 27, 51.

Articles

Allen, F. H., Jr. (1974), Linkage of HL-A and GBG. *Vox Sang.* 27, 382.

Ashall, F., Bramwell, M. E., and Harris, H. (1982), A new marker for human cancer cells. I. The Ca antigen and the Ca antibody. *Lancet* 2, 1.

Baxevanis, G. M., Nagy, Z. A., and Klein, J. (1981), A novel type of T-T cell interaction removes the requirement for I-B region in the H-2 complex. *Proc. Natl. Acad. Sci.* (*USA*) 78, 3809.

Baxevanis, G. N., Wermet, D., Nagy, Z. A., Maurer, P. M., and Klein, J. (1980), Genetic control of T-cell proliferative responses to poly($glu^{60}ala^{40}$) and poly($glu^{51}Lys^{34}tyr^{15}$); subregion specific inhibition of the responses with monoclonal Ia antibodies. *Immunogenetics* 11, 617.

Bergard, I. and Bearn, H. G. (1968), Isolation and properties of a low-molecular-weight β-globulin occurring in human biological fluids. *J. Biol. Chem.* 243, 4095.

Bevan, M. J. (1977), In a radiation chimera, host H-2 antigens determine immune responsiveness of donor cytotoxic cells. *Nature* 269, 417.

Binz, H. and Wigzell, H. (1975), Shared idiotype determinants on B and T lymphocytes reactive against the same antigenic determinants. I. Demonstration of similar or identical idiotypes on IgG molecules and T cell receptors with specificity for the same alloantigens. *J. Exp. Med.* 142, 197.

Bodmer, W. (1972), Evolutionary significance of the HL-A system. *Nature* 237, 139.

Briles, W. E., and Briles, R. W. (1982), Identification of haplotypes of the chicken major histocompatibility complex (B). *Immunogenetics* 15, 449.

Coligan, J. E., Kindt, T. J., Vehara, H., Martinko, J., and Natheson, S. G. (1981), Primary structure of a murine transplantation antigen. *Nature* 291, 35.

Cook, R. G., Vitetta, E. S., Uhr, J. W., and Capra, J. D. (1979), Structural studies on the murine Ia alloantigens. V. Evidence that the structural gene for the I-E/C beta polypeptide is encoded within the I-A subregion. *J. Exp. Med.* 149, 981.

Corte, G., Calabri, R., Damiani, E., Bargellesi, A., Tosi, R., Sorrentino, R. (1981), Human Ia molecules carrying DC1 determinants differ both in α and β subunits from Ia molecules carrying DR determinants. (Lond) 292, 357.

Cullen, S. E., David, C. S., Shreffler, D. C., and Nathenson, S. G. (1974), Membrane molecules determined by the H-2 associated immune response region: isolation and some properties. *Proc. Natl. Acad. Sci.* (*USA*) 71, 648.

Curman, B., Ostberg, L., Sandberg, L., Eriksonn, I. M., Stalenheim, G., Rask, L., and Peterson, P. A. (1975), H-2 linked Ss protein is C4 component of complement. *Nature* 258, 243.

Debre, P., Kapp, J. A., Dorf, M. E., and Benacerraf, B. (1975), Genetic control of specific immune suppression. II. H-2 linked dominant genetic control of immune suppression by the random copolymer L-glutamic acid^{50}-L-tyrosine^{50} (GT). *J. Exp. Med.* 142, 1447.

Demant, P., Capkova, J., Hinzova, E., and Voracova, B. (1973), The role of the histocompatibility-2 linked SS-Slp region in the mouse complement. *Proc. Natl. Acad. Sci.* (*USA*) 70, 863.

Demant, P. and Roos, M. H. (1982), Molecular heterogeneity of D-end products detected by anti-H-2.28 sera. I. A molecule similar to Qa-2, detected in the BALB/cBy but not in the BALB/c.H-2^{dml} mutant. *Immunogenetics* 15, 461.

De Vries, R. R. P., Nijenhuis, L. E., Fat, R. F. M. L. A., and Van Rood, J. J. (1976), HLA-linked genetic control of host response to Mycobacterium leprae. *Lancet* 2, 1328.

Doherty, P. C. and Zinkernagel, R. M. (1975), H-2 compatibility is required for T-cell-mediated lysis of target cells infected with lymphocytic choriomeningitis virus. *J. Exp. Med.* 141, 502.

Dorf, M. E., Maurer, P. H., Merryman, C. F., and Benacerraf, B. (1976), Inclusion group system and cis-trans effects in responses controlled by the two complementing Ir-GL genes. *J. Exp. Med.* 143, 889.

Dorf, M. E., Stimpfling, J. H., and Benacerraf, B (1975), Requirement for two H-2 complex Ir genes for the immune response to the L-Glu-L-Lys-L-Phe terpolymer. *J. Exp. Med.* 141, 1459.

Erb, P. and Feldmann, M. (1975), The role of macrophages in the generation of T helper cells. III. Influence of macrophage-derived factors in helper cell induction. *Eur. J. Immunol.* 5, 759.

Fathman, C. G., Watenabe, T., and Augustin, A. (1978), In vitro secondary MLR. III. Hybrid histocompatibility determinants. *J. Immunol.* 21, 259.

Ferrara, G. B. (1979), Identification of new cell surface markers in man: the problem of immunogenicity. *Transplant. Proc.* 11, 715.

Ferrara, G. B., Tosi, R., Longo, A., Castellani, A., Viviani, C., and Carminati, G. (1978), Safe blood transfusion procedure for immunization against MHC determinants in man. *Transplantation* 26, 150.

Fleischnick, E., Raum, D., Alosco, S. M., Gerald, P. S., Yunis, E. J., Awdeh, Z. L., Granados, J., Crigler, J. F., Giles, C. M., Alpers, A. (1983), Extended MHC haplotypes in 21-hydroxylase-deficiency congenital adrenal hyperplasia: shared genotypes in unrelated patients. *Lancet* 1, 152.

Freed, J. H. and Natheson, S. G. (1977), Similarity of the carbohydrate structures of H-2 and Ia glycoproteins. *J. Immunol.* 119, 477.

Fu, S. M., Kunkel, H. G., Brusman, H. P., Allen, F. H., Jr., and Fotino, M. (1974), Evidence for linkage between HL-A histocompatibility genes and those involved in the synthesis of the second component of complement. *J. Exp. Med.* 140, 1108.

Goldman, M. G. and Goldman, J. N. (1975), Relationship of levels of early components of complement to the H-2 complex of mice. *Fed. Proc.* 34, 979.

Greene, M. I., Pierres, A., Dorf, M. E., and Benacerraf, B. (1977), The I-J subregion codes for determinants on suppressor factor(s) which limit the contact sensitivity response to picryl chloride. *J. Exp. Med.* 146, 293.

Guy, K. and van Heyningen, V. (1982), Further intricacy of HLA-DR antigens. *Immunol. Today* 3, 236.

Jones, P. P., Murphy, D. B., and McDevitt, H. O. (1978), Two-gene control of the expression of murine Ia antigen. *J. Exp. Med.* 148, 925.

Juretic, A., Nagy, Z. A., and Klein, J. (1981), Detection of CML determinants associated with H-2 controlled Eβ and Eα chains. *Nature* 289, 308.

Kapp, J. A., Pierce, C. W., and Benacerraf, B. (1977), Immunosuppressive factor(s) extracted from lymphoid cells of nonresponder mice primed with L-glutamic acid^{60}-L-alanine^{30}-L-tyrosine (GAT). II. Cellular source and effect on responder and nonresponder mice. *J. Exp. Med.* 145, 828.

Katz, D. H., Dorf, M. E., and Benacerraf, B. (1976), Control of T-lymphocyte and B-lymphocyte activation by two complementing Ir-Gl immune response genes. *J. Exp. Med.* 143, 906.

Katz, D. H., Hamaoka, T., and Benacerraf, B. (1973), Cell interactions between histoincompatible T and B lymphocytes. II. Failure of physiologic cooperative interactions between T and B lymphocytes from allogeneic donor strains in humoral response to hapten protein conjugates. *J. Exp. Med.* 137, 1405.

Kaufman, J. F. and Strominger, J. L. (1982), HLA-DR light chain has a polymorphic N-terminal region and a conserved immunoglobulin-like C-terminal region. *Nature* 297, 694.

Kennard, J. and Meruelo, D. (1982), A new murine lymphocyte alloantigen, Ly21.1, mapping to the seventh chromosome. *Immunogenetics* 15, 239.

Kraig, E., Kronenberg, M., Kapp, J. A., Pierce, C. W., Abruzzini, A. F., Sorensen, C. M., Samelson, L. E., Schwartz, R. H., and Hood, L. E. (1983), GAT-specific T and B cells do not transcribe similar heavy chain variable region gene segments. *J. Exp. Med.* 158, 192.

Kronenberg, M., Kraig, E., Siu, G., Kapp, J. A., Kappler, J., Marrack, P., Pierce, C. W., and Hood, L. E. (1983), Three T-cell hybridomas do not contain detectable heavy chain variable gene transcripts. *J. Exp. Med.* 158, 210.

Lieberman, R., Paul, W. E., Humpherey, Jr., W., and Stimpfling, J. H. (1972), H-2-linked immune response (Ir) genes. Independent loci for Ir-IgC and Ir-IgA genes. *J. Exp. Med.* 136, 1231.

Loveland, B. E., McKenzie, I. F. C. (1982), Which T cells cause graft rejection? *Transplantation* 33, 217.

Lozner, E. C., David, H. S., Shearer, G. M., and Terry, W. D. (1979), B-cell alloantigens determined by the H-2 linked Ir region are associated with mixed lymphocyte culture stimulation. *Science* 183, 757.

Mann, D. L., Abelson, L., Harris, S., and Amos, D. B. (1975), Detection of antigens specific for B-lymphoid cultured cell lines with human alloantisera. *J. Exp. Med.* 142, 84.

Mann, D. L., Abelson, L., Henkart, P., Harris, S. D., and Amos, D. B. (1975), Specific human B lymphocyte alloantigens linked to HL-A. *Proc. Natl. Acad. Sci (USA)* 72, 5103.

McKenzie, I. F. C., Morgan, G. M., Sandrin, M. S., Michaelides, M. M., Melvold, R. W., and Kohn, H. I. (1979) B6.C-H-2^{bm12} A new H-2 mutation in the I region in the mouse. *J. Exp. Med.* 150, 1323.

Melchers, I., Rajewsky, K., and Shreffler, D. C. (1973), Ir-LDH$_B$: Map position and functional analysis. *Am. J. Immunol.* 3, 754.

Merryman, C. F. and Maurer, P. H. (1975), Characterization of a new Ir-GLT gene and its location in the I-region of the H-2 complex. *Immunogenetics* 1, 549.

Miller, J. F. A. P., Morahan, G., and Walker, I. D. (1983), T-cell antigen receptors: fact and artifact. *Immunol. Today* 4, 141.

Munro, A. J. and Tussig, M. J. (1975), Two genes in the major histocompatibility complex control immune response. *Nature* 256, 103.

Murphy, D. B., Herzenberg, L. A., Okumura, K., Herzenberg, L. A., and McDevitt, H. O. (1976), A new I subregion (I-J) marked by a locus (Ia-4) controlling surface determinants on suppressor T lymphocytes. *J. Exp. Med.* 144, 699.

Ochi, A., Nonaka, M., Hayakava, K., Okumuro, K., and Tada, T. (1982), Two loci in I-J subregion of the H-2 complex controlling molecules selectively expressed on suppressor and helper T cells. *J. Immunol.* 127, 227.

Ostberg, L., Rask, L., Nilsson, K., and Peterson, P. A. (1975), Independent expression of the two HL-A antigen polypeptide chains. *Eur. J. Immunol.* 5, 462.

Parnes, J. R., Seidman, J. G. (1982), Structure of wild-type and mutant $\beta 2$-microglobulin genes. *Cell* 29, 661.

Parnes, J. R., Velan, B., Felsenfeld, A., Ramanathan, L., Ferrini, U., Appella, E., and Seidman, J. G. (1981), Mouse $\beta 2$-microglobulin cDNA clone: a screening procedure for cDNA clones corresponding to rare mRNAs. *Proc. Natl. Acad. Sci.* (*USA*) 78, 2253.

Peterson, P. A., Cunningham, B. A., Berggard, I., and Edelman, G. M. (1972), $\beta 2$-microglobulin—a free immunoglobulin domain. *Proc. Natl. Acad. Sci.* (*USA*) 69, 1697.

Sachs, D. H. and Cone, J. L. (1973), A mouse B cell alloantigen determined by gene(s) linked to the major histocompatibility complex. *J. Exp. Med.* 138, 1289.

Schrader, J. W., Battye, F., and Scollay, R. (1982), Expression of Thy-1 antigen is not limited to T cells in cultures of mouse hemopoietic cells. *Proc. Natl. Acad. Sci.* 79, 4161.

Schrader, J. W., Cunningham, B. A., and Edelman, G. M. (1975), Functional interactions of viral and histocompatibility antigens at tumor cell surfaces. *Proc. Natl. Acad. Sci.* (*USA*) 12, 5066.

Schwartz, R. H., Dorf, M. E., Benacerraf, B., and Paul, W. E. (1976), The requirement for two complementing Ir-Gl immune response genes in the T-lymphocyte proliferative response to poly(Glu Lys Phe). *J. Exp. Med.* 143, 897.

Schwartz, R. H., Merryman, C. F., and Maurer, P. H. (1979), Gene complementation in the T lymphocyte proliferative response to poly(glu, lys, tyr): evidence for effects of polymer handling and gene dosage. *J. Immunol.* 123, 272.

Silver, L. M. (1982), Genomic analysis of the H-2 complex region associated with mouse t haplotypes. *Cell* 29, 961.

Silver, J., Walner, L. E., Reisfeld, R. A., Pellegrino, M. A., and Ferrone, S. (1979), Structural studies of murine I-E and human DR antigens. *Mol. Immunol.* 16, 37.

Silver, M. and White, M. (1982), A gene product of the mouse t complex with chemical properties of a cell surface-associated component of the extracellular matrix. *Develop. Biol.* 91, 423.

Sorrentino, R., Corte, G., Calabi, F., Tanigaki, N., and Tosi, R. (1982), A three-loci model for the control of the small (beta) subunit of human Ia molecules. In R. Revoltella (Ed.), *Expression of Differentiated Cancer Cells*. New York: Raven Press, p. 69.

Steinmetz, M., Minard, K., Horwath, S., McNicholas, J., Srelinger, J., Wake, C.,

Long, E., Mach, B., and Hood, L. (1982), A molecular map of the immune response region from the major histocompatibility complex of the mouse. *Nature* 300, 35.

Steinmuller, D., Tayler, J. D., Waddick, K. G., and Burlingham, W. T. (1982), Epidermal alloantigen and the survival of mouse skin allografts. *Transplantation* 33, 308.

Strominger, J. L., Orr, H. T., Robb, R. J., Porham, P., Ploegh, H. L., Mann, D. L., Bilofsky, H., Saroff, H. A., Wu, T. T., and Kabat, E. A. (1980), An evaluation of the significance of amino acid sequence homologies in human histocompatibility antigens (HLA-A and HLA-B) with immunoglobulins and other proteins, using relatively short sequences. *Scand. J. Immunol.* 11, 573.

Tanigaki, B. N., Tosi, R. J., Duquesnoy, R. J., and Ferrara, G. B. (1983), Three Ia species with different structures and alloantigenic determinants in an HLA-homozygous cell line. *J. Exp. Med.*, in press.

Tanigaki, N., Tosi, R., Centis, D., Ferrara, G. B., and Pressman, D. (1981), Further characterization of two DR supertypic specificities. Additional evidence that they reside on molecules different from those carrying HLA-DR locus specificities. *Hum. Immunol.* 3, 93.

Tanigaki, N., Tosi, R., Pressman, D., and Ferrara, G. B. (1980), Molecular identification of human Ia antigens coded for by a gene locus closely linked to HLA-DR locus. *Immunogenetics* 10, 151.

Taniguchi, M., Saito, T., and Tada, T. (1979), Antigen-specific suppressive factor produced by a transplacentable I-J bearing T-cell hybridoma. *Nature* 278, 555.

Tosi, R., Tanigaki, N., Centis, D., Ferrara, G. B., and Pressman, D. (1978), Immunological dissection of human Ia molecules. *J. Exp. Med.* 148, 1592.

Tosi, R., Vismara, D., Tanigaki, N., Ferrara, G. B., Cicimarra, F., Buffolano, W., Follo, D., and Auricchio, S. (1983), Evidence that coeliac disease is primarily associated with DC locus allelic specificity. *Clin. Immunol. Immunopathol.* 28, 395.

Tragardh, L., Rask, L., Wiman, K., Fohlman, J., and Peterson, P. A. (1979), Structure and sequence of an immunoglobulin-like HLA antigen heavy chain domain. *Proc. Natl. Acad. Sci.* (*USA*) 76, 5839.

Weiss, E. H., Mellor, A., Golden, L., Fahrner, K., Simpson, E., Hurst, J., and Flavell, R. A. (1983), The structure of a mutant H-2 gene suggests that the generation of polymorphism in H-2 genes may occur by gene conversion-like events. *Nature* 301, 671.

Zinkernagel, R. M. (1976), H-2 compatibility requirement for virus-specific T-cell mediated cytolysis. The H-2K structure involved is coded by a single cistron defined by H-2K mutant mice. *J. Exp. Med.* 143, 437.

Zinkernagel, R. M. and Doherty, P. C. (1974), Restriction of *in vitro* T cell-mediated cytotoxicity in lymphocytic choriomeningitis within a syngeneic or semiallogeneic system. *Nature* 248, 701.

Zinkernagel, R. M., Dunlopp, M. B. C., Blanden, R. V., Doherty, P. C., and Shreffler, D. C. (1976), H-2 compatibility requirement for virus-specific T-cell mediated cytolysis. Evaluation of the role of H-2 I region and non-H-2 genes in regulating immune response. *J. Exp. Med.* 144, 519.

CHAPTER 6

Immunogenetics of the Human Blood Groups

Blood groups are of fundamental importance to immunogenetics for two reasons. First, the original discovery of genetically determined antigenic variations in human red cells was made by Karl Landsteiner, one of the fathers of modern immunochemistry; to our knowledge this was the first area in which genetic principles were found to govern immunological phenomena. Second, blood groups demonstrate several important genetic principles: (a) The site of gene action may not be on the antigenic structure, per se, but rather genes may determine the presence or absence of a given enzyme responsible for synthesis of antigenic moieties, which then become attached to basic precursor substances; (b) gene interaction produces "hybrid" antigens in the offspring of parents both lacking antigens; and (c) antigenic mosaics are formed from individual antigens.

In 1900 Landsteiner obtained blood samples from 22 normal humans, separated the red cells from the serum, and mixed the red cells of each sample with the serum of each of the other. In some cases the blood cells were agglutinated ("clumped"); in others they were not. On the basis of his observations he classified the 22 subjects into three groups. Within each group, blood cells and serum from different subjects could be mixed without producing any agglutination of cells. Further, the blood cells from one group did not clump when mixed with the serum of any other group. In other words they were nonagglutinable. In contrast the blood cells from the other two groups were agglutinated when mixed with the serum of a different group. Dr. Landsteiner accounted for these differences by postulating that blood cells had two kinds of receptors, which he termed A and B; A or B was present in some individuals, whereas in others both A and B properties were absent (group O). He theorized

Table 6.1 Properties of the ABO system

Blood groups	Receptors (antigens) A	B	Agglutinins Anti-A	Anti-B	Approximate incidence (%) in Caucasians
O	−	−	+	+	40
A	+	−	−	+	44
B	−	+	+	−	12
AB	+	+	−	−	4

that the serum proteins from different individuals contain two different substances (agglutinins) that correspond inversely to the A and B properties (receptors) in the blood cells. The agglutinins were capable of reacting with the receptors and thereby agglutinating the cells. The blood serum of any given individual lacks antibody directed against his own blood cells and those of other individuals of the same blood group. The three groups of cells are A (agglutinable by anti-A), B (agglutinable by anti-B), and O (agglutinable by sera from neither A nor B individuals and possessing agglutinators in the serum capable of reacting with both A and B cells). His associates soon thereafter found a fourth group (AB) in which the blood cells contain both receptors (now termed antigens) but lack agglutinins (now termed allohemagglutinins) for both A and B cells (Table 6.1). From this beginning, blood group immunogenetics developed into a major field of investigation. Landsteiner and his associates discovered several other blood group systems, and since then many new blood groups have been found. Over 160 different immunological specificities are now recognized for human red blood cells.

The genetic mechanisms involved in determining the ABO and related (Lewis), I and i blood group antigens differ considerably from the genetics of most other substances. The genetic principles are compatible with simple Mendelian inheritance, but instead of direct production of a given structure (antigen) on red cells, the genes determine the presence or absence of a given enzyme (one or another glycosyl transferase), which is responsible for the attachment of certain sugar residues (which constitute the antigenic determinants) of the ABO and related systems to a set of common precursor glycoproteins (H, I, and i substances). Gene deletions or mutations leading to the absence of precursor H substance result in the rare "Bombay blood" (see p. 186). Cells of individuals with Bombay blood do not reaction with anti-A, anti-B, or anti-H; their serum agglutinates cells of all individuals other than those with Bombay blood, including those of group O.

The biochemical basis for the ABO(H) system has been elucidated in recent years (see p. 190). Since the structural basis of most other blood group substances is still unknown, the genetic mechanisms involved remain obscure. If they prove to be proteins, the genes that determine the Rh_o antigen, for instance, may act directly to produce the Rh_o substance within the cell. Should they not be of protein nature, one would suppose that the genes would instead involve the production of specific enzymes that attach certain specific moieties to a precursor substance, resulting in an antigen detectable by current serological means.

With three exceptions, human red blood cell antigen systems discovered to date are inherited as simple autosomal Mendelian traits. The exceptions are (1) the so-called Xg^a antigen, which is sex-linked, (2) the Lewis antigens Le^a and Le^b, whose genetics are very complex and illustrate an important genetic principle, "hybrid antigens," and (3) the Ii system, whose genetics are unclear.

BLOOD GROUP METHODOLOGY

Blood group antibodies can be "immune" (formed as a result of exposure to foreign antigens, as in transfusion or pregnancy) or "naturally occurring" (present without any known antigenic stimulus). Antibodies that agglutinate saline-suspended red cells are termed "complete" antibodies; those antibodies to blood group antigens that do not agglutinate cells suspended in saline are termed "incomplete." The so-called naturally occurring antibodies are usually IgM "complete" antibodies (for example, anti-A, -B, -H, -I, -M, -N, -P, -Lewis) and are easily detected by simple agglutination of saline-suspended red cells. The "immune" antibodies are usually IgG. IgG antibodies, except for anti-A and anti-B, are usually "incomplete" antibodies and do not directly agglutinate saline-suspended cells; it has been suggested that the IgG molecule is not large enough to "bridge the gap" between adjacent red cells, in contrast to the larger IgM molecule. The red cells can be brought close enough together to allow IgG molecules to agglutinate by altering the zeta potential. The zeta potential is the determinant factor of the repulsive effort exerted in maintaining two cells a given distance apart in any suspending medium. This can be changed by (1) altering the dielectric constant (the charge-dissipating power) of the suspending medium by adding substances such as bovine albumin or (2) altering the charge of the red cell membranes by treating them with proteolytic enzymes such as trypsin, papain, ficin, and bromelin.

Another way of detecting IgG incomplete antibodies is by means of the antiglobulin (Coombs) test. Blood group antibodies added to the red cells that combine with them without producing agglutination are detected by subsequent washing of the mixture of coated red cells and serum to remove all serum proteins, followed by addition of a rabbit antiserum to human globulin; the latter reacts with the human globulin on the cells and causes agglutination (Fig. 6.1).

Recently, the reactions of IgG and IgM anti-A and anti-B antibodies with A and B blood group glycoproteins have been examined using a radioimmunoassay system (Farr technique). The resulting inhibition curves indicated significantly higher binding constants for the IgM antibodies (possibly due to its pentravalence) than for the bivalent IgG, and it may be that this difference accounts for the fact

Fig. 6.1 "Complete" and "incomplete" antibodies.

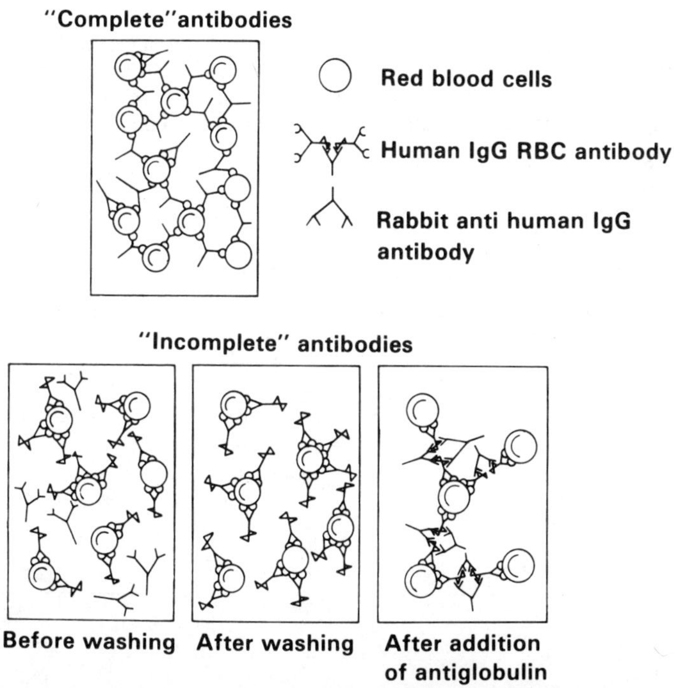

that A and B glycoproteins readily inhibit agglutination by IgG anti-A and anti-B.

STEPS IN THE DELINEATION OF A NEW BLOOD GROUP SYSTEM

For the recognition of most new blood group systems, the antiserum defining the "new" antibody is first tested against a panel of cells similar to that shown in Table 6.2. If the pattern of reactivity does not correspond to any known antibody (or mixture of antibodies), the antiserum presumably identifies a new antigen, here hypothetically termed Z. Family studies are then performed to ascertain (1) whether Z is inherited in accordance with simple Mendelian genetics and (2) whether the antigen segregates independently in families of all hitherto unknown blood group systems. Cells failing to react with the antibody, for example, Z negative, presumably contain the alternative allele, *z*. The phenotype frequency of Z+ and Z— is established by population studies. When another new antibody is subsequently found that reacts with all Z— cells, the z antigen is now detectable. Population studies are then performed to see if the incidence of ZZ, Z*z*, and *zz* cells is in keeping with that ex-

Table 6.2 A typical cell panel containing 10 red cell specimens with different antigenic compositions—e.g., cell 4 lacks the high-incidence antigen Cartwright (YC^a). Specificities of antibodies in a serum or eluate can be determined from its reaction pattern with the cells. For example, positive reac-

	Blood group system		Rh–hr									MNS					
No.	Special type	Rh–hr code	D	C	Cw	E	c	e	f	V	VS	T_M	M_1	M	N	S	s
1		r′r′-1	o	+	o	o	o	+	o	o	o	+	o	+	+	+	+
2		$R_1R_1^w$-51	+	+	+	o	o	+	o	o	o	o	o	+	+	o	+
3		R_1R_1-86	+	+	o	o	o	+	o	o	/	/	o	+	o	o	+
4	Yt(a—)	R_2R_2-62	+	o	o	+	+	o	o	o	/	o	o	+	+	+	+
5		R_2R_2-6	+	o	o	+	+	o	o	o	o	+	o	o	+	+	+
6	DBG+	r″r″-2	o	o	o	+	+	o	o	o	/	o	o	+	o	o	+
7		rr-78	o	o	o	o	+	+	+	o	o	o	o	+	+	o	+
8	Kp(a+)	rr-80	o	o	o	o	+	+	+	o	/	o	o	o	+	+	o
9		rr-90	o	o	o	o	+	+	+	o	/	/	o	o	+	o	+
* 10		R_0r-10	+	o	o	o	+	+	+	+	/	/	o	+	+	o	+

pected. Usually Z and *z* will be codominant; that is, the heterozygote will be recognized by both anti-Z and anti-z antisera. If frequencies of ZZ, Z*z*, and *zz* do not total 100%, then one or more additional alleles can be expected. Future studies may indicate that the new system is closely associated with other allelic genes. With some antisera and antigens a "dosage effect" is observed; that is, zz (homozygotes) given stronger reactions than Z*z* (heterozygotes) with anti-z.

ABO BLOOD GROUPS

Following Landsteiner's original experiments, Bernstein suggested that the ABO blood groups were inherited in a Mendelian manner. It has now been shown, using somatic-cell hybridization techniques, that the *ABO* genes are linked to the red cell adenylate kinase-1 (AK_1) and the nail–patella syndrome (N_p) located on chromosome 9. He postulated three allelic genes, *A*, *B*, and *O*. A chromosome carrying *A*, *B*, or *O* is inherited from each parent. The six possible genotypes are therefore *AA*, *AO*, *BB*, *BO*, *AB*, and *OO*. As the only available antisera are anti-A and anti-B, only four phenotypes, A, B, AB, and O, are determined in the laboratory, *AA* being indistinguishable from *AO* and *BB* from *BO*. For many years confusion

tions with cells 4 through 10 and negative reactions with cells 1, 2, and 3 indicate reactivity only with cells possessing the c antigen. Thus the specificity of the antibody would be anti-c (anti-hr′).

Lutheran (Lu)		P	Lewis (Le)		Kell					Duffy (Fy)		Kidd (Jk)		Sex Link	Dombrock
a	b	P_1	a	b	K	k	Kw	Js^a	Js^b	a	b	a	b	$X_g{}^a$	Do^a
o	+	+	o	+	o	+	o	o	+	+	o	o	+	+	/
o	+	+	o	+	o	+	o	o	+	+	+	+	+	+	+
o	+	o	o	o	o	+	o	o	+	+	+	+	+	+	/
o	+	+	+	o	+	+	/	o	+	+	+	+	+	+	/
o	+	$+^w$	o	+	+	+	o	o	+	o	+	+	+	+	/
o	+	+	o	+	o	+	o	o	+	o	+	o	+	+	/
o	+	+	o	+	o	+	o	+	+	o	o	+	+	+	/
o	+	+	+	o	o	+	o	o	+	+	+	+	+	+	/
+	+	+	+	o	+	+	/	o	+	+	+	+	o	+	/
o	+	+	+	o	o	o	o	o	+	+	o	o	+	o	/

existed regarding the O gene, for some animal and human sera were found that reacted more strongly with O cells than A or B and were incorrectly called anti-O. These sera could not, however, differentiate between AA and AO individuals; moreover many of them were inhibited by saliva from type AB individuals, who could not secrete a product of the O gene. It is convenient to talk in terms of an O gene, but it is in fact an amorph, that is, a gene exerting no demonstrable effect. It was shown that all normal individuals possess a precursor blood group substance that is acted on by a gene called *H* and converted to H substance. This in turn is acted on by the respective A, *B*, and O genes. The A and *B* genes convert the H substance to A or B substances, leaving a variable amount of H substance unconverted. Where no *A* or *B* genes are inherited (type O), the H substance is totally unconverted.

Precursor substance —*H* gene→*H* substance
- ↗ A gene—A antigen and some H antigen
- → *B* gene—B antigen and some H antigen
- ↘ O gene—H antigen

These simple principles have now been extended by immunochemical studies and are discussed later in this chapter.

Common Subgroups of A

The usual anti-A serum distinguishes two different kinds of A cells, A_1 and A_2, only by titration score, but the two can be clearly differentiated by certain lectins (plant extracts),* for example, from *Dolichus biflorus*. With these, A_2 reacts more weakly than A_1 with anti-A. The differences between A_1 and A_2 are genetically determined, but the biochemical basis for this difference is unknown; whether the difference is quantitative or qualitative is a matter of considerable controversy. Anti-A_1, detectable at room temperature, occurs in the sera of 1 to 2% of A_2 people and in about 25% of A_2B people. A few samples not classifiable either as A_1 or A_2 are called intermediates. Table 6.3 lists the common ABO genotypes and phenotypes.

* Certain plant extracts ("lectins") react specifically with H substance, that is, possess anti-H specificity. Many postulates for the presence of such serological activity in plant extracts have been put forth, but none is universally accepted. Anti-A_1 can be produced by absorption of anti-A serum by A_2 cells.

Table 6.3 The common ABO phenotypes and genotypes

Phenotypes	Genotypes
A_1	A_1A_1, A_1O, A_1A_2
A_2	A_2A_2, A_2O
B	BB, BO
O	OO
A_1B	A_1B
A_2B	A_2B

Rare Subgroups of ABO

A_3, first discovered by Friedenreich in 1936, is characterized by "mixed-field agglutination"; addition of anti-A serum or of sera from group O individuals (anti-A plus anti-B) produces very small clumps of agglutinated cells and much larger numbers of unagglutinated cells. However, certain group O sera (Dunsford) can agglutinate all the cells of people of genotype A_3. The A_3 characteristic is genetically inherited, and many cells of apparent A_3 genotype in fact arise by modifying interaction between A_2 and *B* genes. A_3 must be distinguished from chimeras (mixtures of two blood types present in some twins), from the effects of bone marrow transplantation or transplacental hemorrhage, and from changes in antigenic strength that sometimes occur in leukemia and other diseases and sometimes in normal persons for no known reason.

Other weak forms of the A antigen exist. These are of two main groups:

1. A_x (A_x, A_4, A_5, A_o, A_{e1}). Cells of such individuals give negative or weak reactions with anti-A from B donors but good reactions with anti-A from O donors. The serum of such individuals agglutinates A_1 cells but not A_2. The saliva of secretor individuals of this type contains H substance but usually no A substance.
2. A_m type. These cells react weakly or not at all with the anti-A from either B individuals or O individuals. The serum lacks antibodies to both A_1 and A_2. The saliva (if secretor) contains both A and H.

Hereditary weak A and B blood group variants are characterized by extremely weak expression of these antigens in red cells, with normal expression of H antigen. From classical immunogenetic data, many phenotypic variations are distinguishable, including A_3, A_x, A_{end}, A_m, and A_{el}. The number of A receptors per cell ranges from 3,000 to 140,000 depending on the variation. It would be important to characterize the antisera detecting these "weak" variants as to IgM or IgG content, as these show different abilities to hemagglutinate. Similar terminology is used in practice to classify weak B variants (B_3, B_x, B_{el}, etc.).

Quantitative methods have provided important contributions to our understanding of the ABO system. For example, in the weak A phenotypes the number of reaction sites on the red cell varies from more than 800,000 for A_1 to roughly 5,000 for A_x and still fewer for A_m, A_{end}, and A_{el}. In the case of A_{end}, Cartron has shown that the agglutinable population has about as many sites as found on A_2 cells, whereas the number of sites in free cells is practically zero, as if this phenotype were actually a mixture of A_2 and O. He has also shown, using prolonged incubation of the serum in the presence of labeled sugar, that the *N*-acetylgalactosylaminyltransferase enzymes in the weak A_m phenotype exist in two forms, corresponding to A_1 and A_2, suggesting that there are in fact two kinds of A_m—A_m^{a1} and A_m^{a2}. These characteristics of the primary gene products were proved to be useful in determining the genotype, and it thus became possible for the first time in the ABO system to detect a recessive character at the level of the primary gene product.

Indeed, studies of the glycosyl transferases provide a useful approach to the ABO system, and particularly the *ABO* locus structure and function. The extreme variability of the ABO gene products suggests heterogeneity of the locus. For example, a specific *N*-acetylgalactosaminyltransferase has been shown to be present in an "enhanced" A_2B subject but absent in A_x individuals from the same family, providing the first evidence that genetic complementation may occur within the *ABO* locus. An analysis of blood group gene-specified glycosyl transferases has recently led to the recognition of a human dispermic chimera. These individuals are believed to arise from the fertilization of two maternal nuclei by two sperm and the fusion of the two zygotes to form one organism.

Anomalies of the ABO System

The "Bombay" phenotype. This phenotype is a rare phenomenon in which the patient's cells do not react with anti-A, anti-B, or anti-H. This serum contains anti-A or anti-B and a powerful anti-H. These cells have been designated Oh because they appear in the laboratory to be type O but lack H substance. It was originally suggested that this can be explained by the effect of a suppressor gene *h*. Normal subjects possess a modifying gene *H* (*HH* or *Hh*) responsible for the development of H substance. The Bombay blood propositus inherits a rare double dose of the suppressor gene (*hh*), which prevents the development of the H antigen on the red cell and in the secretions. The B blood group gene is inherited normally, but because there is no H substance for it to act on, the B antigen is not formed. Obligate heterozygotes are shown to have normal blood groups, since they inherited the *A* or *B* gene from the propositus and a normal *H* gene from the other parent. Since they are genotype *Hh*, they are able to express the *H* gene normally. Oh can also be the manifestation of inhibition of A_1 and A_2 as well as of *B* and *H*. These phenotypes are designated Oh^B, Oh^{A1}, and Oh^{A2}. It is important to mention that blood group H substance can be isolated from the red blood cell membrane of a "Bombay" Oh subject who synthesized anti-H, which was neutralized by his own H substance *in vitro*. This indicates that the immune system did not have access to the concealed H determinants that had been present in the red blood cells since embryonic life.

Acquired B antigen. Some group A patients have an acquired rather than an inherited B antigen, often associated with carcinoma of the colon and recently with a *Proteus vulgaris* septicemia. It has been known for some time that certain organisms (e.g., *E. coli* and *Proteus*) have B-like activity. Purified lipopolysaccharides from such organisms have been shown to adsorb to red cells of type A and O *in vitro*, giving them a B-like antigen. The exact mechanism for the acquisition *in vivo* is not proved but is thought to be due either to absorption of bacterial polysaccharide by the red cells or to alteration of the red cell membrane by bacterial enzymes. Such "acquired" B antigens are found rarely in normal A individuals without infection.

Weak A and B antigens. Weak variants of B are rarer than variants of A but have been described. Apart from the weak subgroups of A described earlier, which are inherited, there are other weakenings seen due to disease (e.g., in leukemia) and due to gene interaction (for example, the A_2 antigen of an AB can appear as a weak

A_3 antigen). The laboratory results obtained when typing patients with such weak antigens—or also with human blood group chimeras, where perhaps only a minor population of A cells are present—must be interpreted with caution in conjunction with family studies.

Origin of "Natural Isohemagglutinins"

The origin of anti-A and anti-B in individuals lacking these antigens probably reflects cross-reactivity of these antigens with various bacterial antigens. Thus, germ-free chickens lack anti-B, but when exposed to bacteria containing antigens cross-reacting with B substance, they develop anti-B. The anti-A and anti-B titers present in normal individuals ("normal isoagglutinin") can be boosted deliberately by transfusion of incompatible blood, by transfusion of incompatible secretor plasma, by injection of cross-reacting substances (gastric mucosa is a good source of A antigen, horse mucosa of B antigen), or by injection of certain influenza virus vaccines.*

Pregnancy with an ABO-incompatible fetus may also result in an increase in titers of these isoagglutinins ("hyperimmune isoagglutinin"). In general, normal isoagglutinins are IgM and "hyperimmune" IgG. (In group O individuals, some IgG anti-A and anti-B are usually present, as are IgA antibodies of the same specificities.) A- and B-like substances and materials that react with antibodies thereto appear ubiquitous in nature. Some plant extracts ("lectins") will agglutinate A_1 but not A_2 cells (e.g., *Dolichus biflorus*) and will also precipitate the saliva of secretors of group A_1 but not A_2. Other lectins agglutinate only B cells. Still others, especially *Ulex europeus*, agglutinate all cells containing H substance (i.e., all except Bombay bloods). Substances with various specificities for A_1, A_2, B, and O cells have also been found in various species of snails. Investigations of these agglutinins are currently in progress in many laboratories.

Secretor and Nonsecretor

About 80% of the population secrete the soluble blood group substances with A, B, and O(H) specificities corresponding to each

* This is due partly to A-like substances in chick embryos and partly to A-like substances in the fluids in which the virus particles are suspended. The cytoplasmic substances of the host (chick embryo) become incorporated into the virus.

individual's blood type;* the remaining 20%, termed "nonsecretors," do not. Thus a type A "secretor" has A substance present in body fluids such as saliva, gastric juice, sera, and tears, and if a man, in semen, and if a woman, in breast milk. Secretion, and nonsecretion, of A and B and H substances is controlled by a pair of allelic genes, *Se* and *se*. Homozygotes for *se* are nonsecretors. Most nonsecretors do, however, secrete large amounts of Lewis(a) blood group substance, a glycoprotein similar in composition to the A, B, and O(H) substances. On the basis of chemical analysis of A, B, and H substances, Watkins and Morgan proposed that the absence or marked diminution of the enzyme produced by the *H* genes in the secretory organs of the nonsecretors results in their inability to synthesize glycoproteins with A, B, or O structures. Moreover, Watkins also proposed that differences between A and B individuals and Le(a+) and Le(a−) individuals result from the genetically determined presence or absence of other specific transglycosylases (see below).

The Lewis Blood Group System

Anti-Lewis(a), or anti-Le^a, was first described by Mourant in 1946; the antibody reacted with the red cells of 22% of the population tested. In 1948 Andresen described an antibody giving antithetical results, anti-Le^b. Later in that year Grubb made the observation that individuals whose red cells were Le^{a+} were nonsecretors of ABH substances. In 1951 he also described the presence of Le^a and Le^b substances in saliva. Other workers subsequently showed that Le^a and Le^b substances were also present in serum and that these substances could be transferred to red cells lacking these antigens by simple incubation. It was postulated that the Lewis antigens were in fact primarily a system of antigens present in body fluids and only secondarily on red cells. These data, together with the genetic studies of Ceppellini and the chemical studies of Watkins and Morgan, led to the theory of the Lewis system accepted today. It is postulated that a precursor blood group substance is converted through a series of reactions to the specific blood group substances appearing on the red cells and in the body fluids. These reactions are controlled by

* Other blood group substances have been found in body fluids, for example, Sd^a, Lewis, and I, but are under different genetic control from that controlling secretion of ABO(H) substances. Further, Sd^a appears to be present in varying concentrations in all normal individuals.

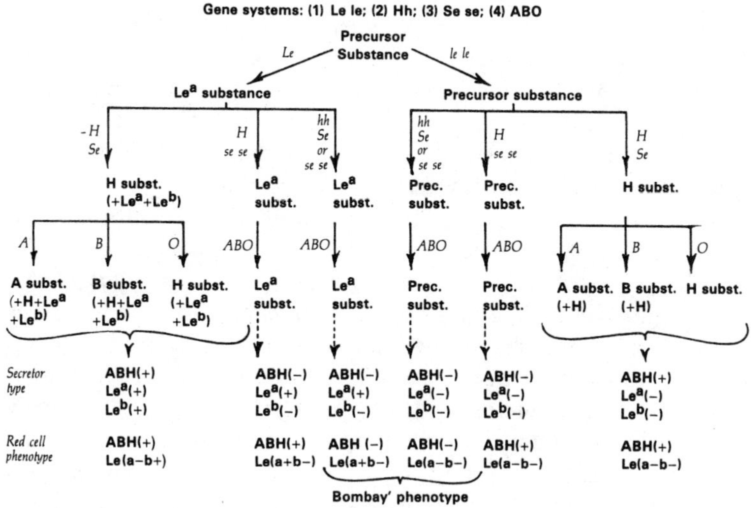

Fig. 6.2 Possible genetic pathways for the biosynthesis of A, B, H, and Le^a substances. (From W. M. Watkins, Biochemistry and genetics of the ABO, Lewis, and P blood group systems, *Advances in Human Genetics* 10: 1, 1980)

four sets of genes, all of which are inherited independently, acting in sequence: first the *H* genes (*H* and *h*), second the *A* or *B* genes, third the secretor genes (*Se* and *se*), and last the *Le* gene (see Fig. 6.2).

Boettcher has suggested that the secretor gene exerts its influence in the appropriate tissue before *H* and *ABO* genes. This is based on the presence in secretor serum of a specific fucosyl transferase that attaches L-fucose by a 1→2 linkage to terminal nonreducing galactosyl residues of oligosaccharides. This does not occur in nonsecretors. Interestingly, it has recently been proposed that *H* and *Se* are two closely linked structural genes that code for two different 2-α-L-fucosyl transferases. The biochemical basis for this is explained in the section that follows.

*Immunochemistry of ABH and Lewis (Le) Substances**

The soluble antigens, A, B, and H, being available in plentiful supply in secretions such as ovarian cyst fluids, have readily lent them-

* Abbreviations used include: Gal for galactose, Fuc for fucose, GlcNAc for *N*-acetylglucosamine, and GalNAc for *N*-acetylgalactosamine. All monosaccharides except fucose have a D configuration; fucose has an L configuration.

selves to characterization by chemical purification. Extensive studies by Watkins, Morgan, and Kabat have shown that all three antigenic determinants of the ABO system (A, B, and H) and of the Lewis substances, Le^a and Le^b, have essentially the same chemical composition, namely, glycoprotein molecules containing numerous oligosaccharide chains composed of five sugars: D-galactose, L-fucose, *N*-acetyl-D-glucosamine, *N*-acetyl-D-galactosamine, and sialic acid (Fig. 6.3). The oligosaccharide chains are attached to a peptide "backbone" that governs molecular conformation. The differences in serological specificity of A, B, H, and Le^b are due to the terminal

Fig. 6.3 Structures of the five principal sugars providing the backbone of ABH and Lewis substances. The first four are involved in the antigenic determinants attached glycosidically to the oligosaccharides and are responsible for specificity. (a) *N*-acetyl-D-galactosamine; (b) D-galactose; (c) L-fucose; (d) *N*-acetyl-D-glucosamine; (e) *N*-acetylneuraminic acid. (From W. M. Watkins, Blood groups substances, *Science* 152, 175, April 8, 1966. Copyright © 1966 by the American Association for the Advancement of Science. By permission of the author and publisher.)

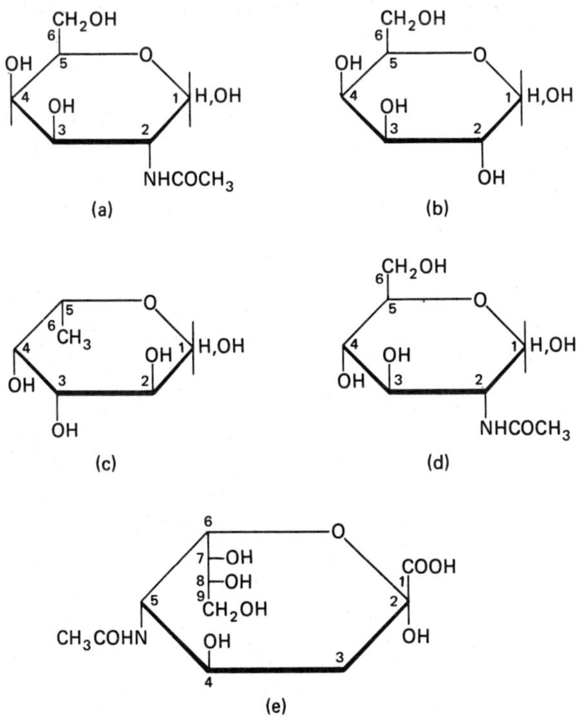

monosaccharide and its linkage. Recent work shows that ABO(H) antigens extracted from the red cells have essentially similar carbohydrate structure but appear to be exclusively glycolipid. There are two distinct types of carbohydrate chain endings that appear to form the basis of the A, B, and H determinants. The Type I chain ending has a β-galactosyl residue joined by 1→3 linkage to *N*-acetylglucosamine. The Type II ending has a 1→4 linkage between these two sugars. Regardless of the length and complexity of the carbohydrate chains, the erythrocyte A, B, and H determinants at the end of the chains are identical with the secreted ovarian cyst fluid glycoproteins—except that the secreted glycoprotein active moieties are found to be carried on both Types I and II chains, whereas, in the glycolipid isolated from the red cell membranes, the moieties are carried only on the Type II chain endings. Figures 6.4 and 6.5 depict the sugar residues involved in A, B, H, and Lewis specificities.

Biochemical data accumulated since the mid-1960s indicate that the inheritance of the ABO(H) and Lewis types is very complex. Family studies show that the inheritance of these determinants is controlled by genes at four independent loci, namely, the *ABO*, *Lele*, *Hh*, and *Sese* loci. The *A* gene produces an *N*-acetylgalactosaminyl transferase, the *B* gene a galactosyl transferase, and the *H* and *Le* genes produce two different fucosyl transferases. The *Se* gene does not produce an enzyme but is necessary for the formation of the fucosyl transferase specified by the *H* gene in certain organs and not in others. Its presence does not control the expression of the *H* gene, and its presence or absence does not affect the formation of A, B, or H determinants on red cells. The *Se* gene does, however, affect Lewis blood types. Lewis substances are not integral parts of the red cell membrane but rather are sphingoglycolipids adsorbed from the serum onto the red cells. In the serum they circulate with the serum lipoproteins; they are not present on umbilical cord blood cells but become detectable at 3 to 6 months of life, reaching adult concentrations at about 6 years of age. The clear association between secretor status and Lewis blood type was first pointed out by Grubb in 1951: all Le(a−b+) individuals are secretors of ABH substance, having ABH substances in saliva, gastric juice, milk, and so forth, detectable by appropriate antisera and standard serological techniques, such as inhibition of agglutination; Le(a+) individuals are always nonsecretors of ABH substances; Le(b−) individuals are secretors; and Le(a−b−) may be either secretors or nonsecretors.

As previously noted, the two closely similar backbones of blood

(a)

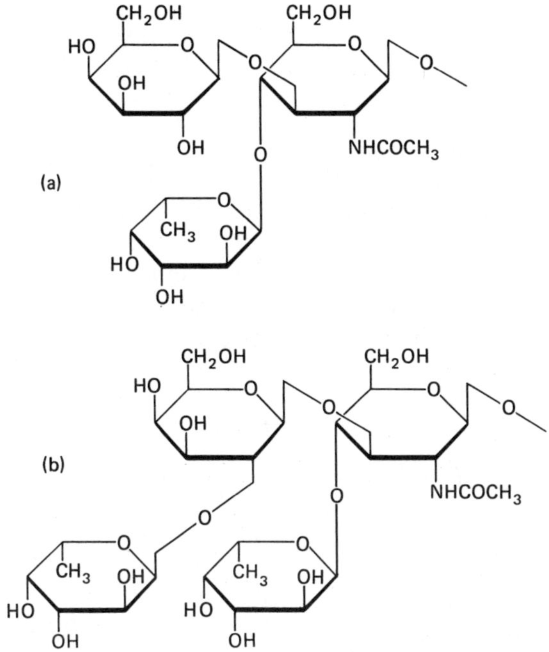

(b)

Fig. 6.4 Sugar residues involved in Lewis antigenic determinants. (a) Le^{a}-active substance. (b) Le^{b}-active substance. (From W. M. Watkins, Blood groups substances, *Science* 152, 175, April 8, 1966. Copyright © 1966 by the American Association for the Advancement of Science. By permission of the author and publisher.)

group oligosaccharides both contain the disaccharide unit β-galactosyl-*N*-acetylglucosamine, with the two sugars linked either 1→3 (Type I) or 1→4 (Type II). If the precursor substance contains both types, additions to these precursor disaccharide structures form the basis of the A-, B-, H-, Le^{a}, and Le^{b} specific substances.

The proposed additions to these two types of chain endings controlled by the *H*, *Le*, *A*, and *B* genes and the serological specificity of the structure formed are summarized in Table 6.4. Thus the *H* gene is considered to control the addition of L-fucose to the terminal nonreducing galactosyl residue of the Type I or Type II chain in α(1→2) linkage to give H-specific structures. The *Le* gene controls the addition of L-fucose to the subterminal *N*-acetylglucosamine residue in α(→4) linkage in the Type I chain to give an Le^{a}-active structure. The antigenic determinant for the Le^{a} specificity of soluble blood group substances has the structure

L-Fuc
|
α(1→4)
|
D-Gal—β(1→3)—D-GlcNac—β(1→3)—R

Such an addition cannot take place on the Type II chain, because the 4-position of *N*-acetlyglucosamine is already substituted with the galactosyl residue. When both the *H* and *Le* genes are present, substitutions with L-fucose occur on both the galactosyl and the *N*-acetylglucosaminyl residues and give rise to the structure with Le^b specificity:

L-Fuc   L-Fuc
| |
α(1→2)   α(1→4)
| |
D-Gal—β(1→3)—D-GlcNac—β(1→3)—R

(a)
(b)

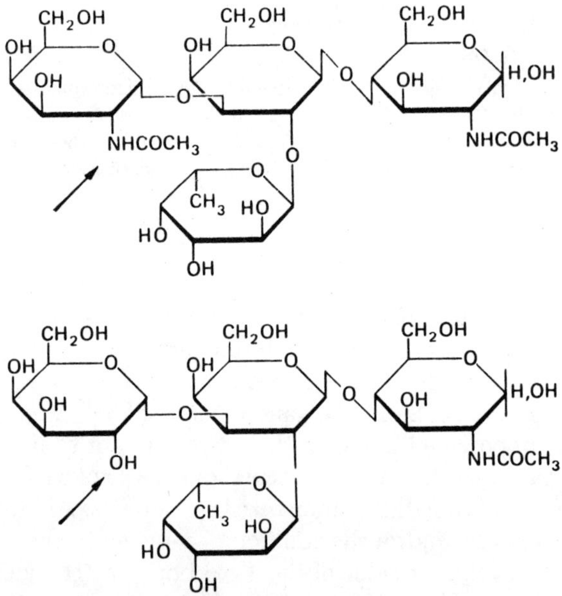

Fig. 6.5 Serologically active tetrasaccharides derived from alkaline degradation product of human (a) A substance (b) B substance. (From W. M. Watkins, Blood groups substances, *Science* 152, 176, April 8, 1966. Copyright © 1966 by the American Association for the Advancement of Science. By permission of the author and publisher.)

The addition of this second L-fucose is catalyzed by a specific L-fucosyl transferase present in milk from secretors of soluble blood group substances A, B, or O(H) corresponding to their particular ABO blood type and is absent from milk of nonsecretors. Both Le^a and Le^b determinants contain L-fucose in $\alpha(1 \rightarrow 4)$ linkage to *N*-acetyl-D-glucosamine.

When both *H* and *A* genes are present, the precursor chains are converted first to H-active structures by the addition of L-fucose in $\alpha(1 \rightarrow 2)$ linkage, and then to A-specific groupings by the substitution of the terminal β-galactosyl residue at the 3-position with an α-*N*-acetylgalactosaminyl unit. When both *H* and *B* genes are present, the β-galactosyl unit in the H-active structures is substituted, also in the 3-position, with an α-D-galactosyl unit to give a B-active structure. Thus the difference between A- and B-antigen activity is solely due to the substituent group attached to the second carbon atom of a galactose molecule: an acetyl amino group for A sub-

Table 6.4 Genes, gene products, and antigenic structures responsible for ABH and Lewis specificities

Gene	Gene product	Substance transferred	Antigenic structure
H(O)	Fucosyl transferase	α-L-Fucose	Galβ1-3GlcNAc 2 or 4 | Fucα1
A	N-Acetylgalactosamine transferase	α-D-N-Acetyl-galactosamine	GalNAcα1-3Galβ1-3GlcNAc 2 or 4 | Fucα1
B	Galactosyl transferase	α-D-Galactose	Galα1-3Galβ1-3GlcNAc 2 or 4 | Fucα1
(*Le^a*)	Fucosyl transferase	α-L-Fucose	Galβ1-3GlcNAc 4 | Fucα1
(*Le^b*)	Fucosyl transferase (presence of H fucosyl transferase of secretors is required for this gene to act)	α-L-Fucose	Galβ1-3GlcNAc 2 4 | | Fucα1 Fucα1

stance and an hydroxyl group for B substance. The *N*-acetylgalactosamine or galactose residues mask the H specificity of the structure to which they are added, and indeed the A and B serological determinants may be looked on as interaction products of the *A* and *H* and the *B* and *H* genes, respectively.

The inner sugars to which these are attached (designated "R" above) were first found to be glycoproteins in the secreted blood group entities (Lloyd and Kabat, 1968) and shown to be glycolipids on the red cells (Hakomori, 1970, 1974), with the following general structure:

(attachment
site)
↓
. . . —D-GlcNAc—β(1→3)—D-Gal—β(1→4)—Glu—Ceramide.

The Lewis system is defined by two allelic genes, *Le* and *le*, which are inherited independently of the *ABO*, *Hh*, and *Sese* genes. The change produced by the *Le* gene occurs on the same macromolecules as those produced by *Hh* and *ABO* genes. In single or double dose *Le* gives rise to Le^a specific structures; absence of an *Le* gene (i.e., presence of *lele*) results in the absence of Le^a- and Le^b-specific structures. The Le^b specificity is now thought to be an interaction product of the *H* and *Le* genes.

The Le^a and Le^b substances are not intrinsic to the red cells but are adsorbed to them. Le^a is present at birth but not attached to the red blood cells. Babies with the *Le* and *Se* genes are first Le(a+b−), then Le(a+b+), then Le(a−b+) during a period of 1 to 2 years. Full expression is achieved only at about 3 to 6 years of age.

Six red cell phenotypes can be distinguished on the basis of ABO and Lewis phenotypes on the red cells and the presence of A, B, H, Le^a, and Le^b substances in secretions (Table 6.5). The first four types are common. Groups 5 and 6 correspond to individuals of the very rare Bombay phenotype mentioned earlier, for example, homozygous *hh* individuals.*

The structures responsible for the ABH and Lewis specificities were shown in Figures 6.4 and 6.5. These are attached to an inner

* By family studies Bombay individuals can be shown to possess an *A* gene or *B* gene. Such individuals are termed O_h^B and O_h^A, respectively.

Table 6.5 The gene combinations giving rise to A, B, H, Le^a, and Le^b activities on the red cell and in secretions

	Antigens on red cells					Substances in secretions					
Gene combinations	A	B	H	Le^a	Le^b	A	B	H	Le^a	Le^b	
ABO,H,Se,Le	+	+	+	−	+	+	+	+	+	+	
ABO,H,sese,Le	+	+	+	+	−	−	−	−	+	−	
ABO,H,Se,lele	+	+	+	−	−	+	+	+	−	−	
ABO,H,sese,lele	+	+	+	−	−	−	−	−	−	−	
ABO,hh,Se or *sese,Le*	−	−	−	+	−	−	−	−	+	−	Bombay phenotype (rare)
ABO,hh,Se or *sese,lele*	−	−	−	−	−	−	−	−	−	−	

branched oligosaccharide common to all specificities, shown above. The enzymes responsible for the attachment are also listed in Table 6.4. The two kinds of carbohydrate chains in human A, B, H, and Le substances differ only in the linkage (whether 1→3 or 1→4) of the subterminal galactose to the next sugar in the chain, *N*-acetylglucosamine. The ABH and Lewis antigenic determinants are *secondary* gene products, the *primary* gene products being the enzymes responsible for the attachment of the appropriate sugar moiety. These enzymes, working in sequence, determine the specificity formed (see Fig. 6.6). This mechanism of synthesis has important genetic implications in that it provides a biochemical basis for antigens produced by "gene interaction," that is, antigens such as the Le^b determinant present in a hybrid that are not found in either parent. It now appears that the Le^b antigen is an interaction product of three genes, namely, the *H*, *Le*, and *Se* genes. *Le* and *Se* specify the two fucosyl transferases required for the synthesis of the Le^b determinant. Hence a child inheriting an *Le* gene from an *Le*/*se,se* parent and an *Se* gene from an *le,le*/*Se* parent can synthesize a structure that neither parent can make.

Recent biochemical studies have produced evidence for the elucidation of the genetic mechanisms described above. Some of the more important studies are summarized below.

Incorporation of labeled L-fucose from guanosine diphosphate L-fucose into endogenous H-active material in hog gastric mucosa was demonstrated by Grollman and Marcus (1966), and the product was characterized by immunochemical methods. These experi-

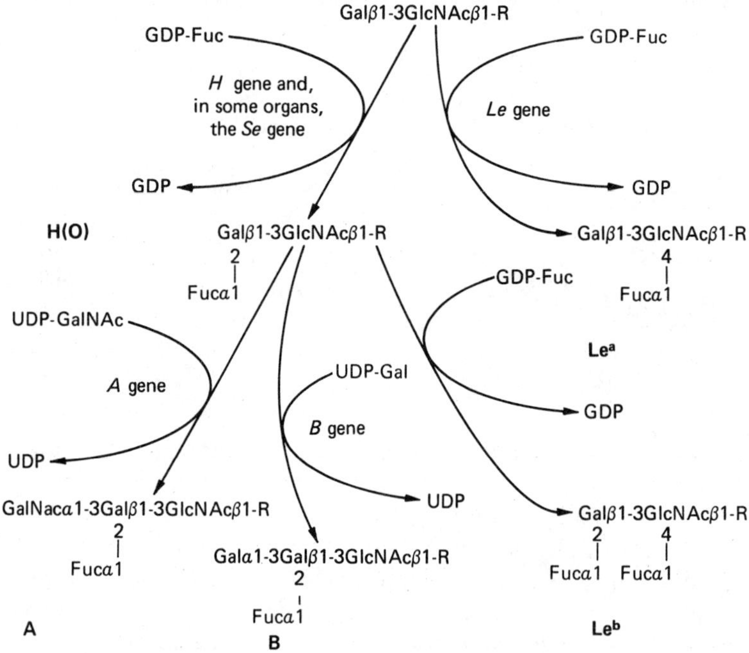

Fig. 6.6 Glycosyl transferases specified by the genes that determine blood type. (From V. Ginsburg *et al.*, in *Glycoproteins of Blood Cells and Plasma,* G. A. Jamieson and J. Tibor, Eds., Philadelphia: Lippincott, 1971, by permission of the author and publisher.)

ments did not, however, establish the site of attachment of the fucose or the mechanism by which the sugar was incorporated. The demonstration of the enzymes with the specificities expected from the product of the A and B genes nevertheless provided tools for the further investigation of the problem of macromolecular biosynthesis.

There is still substantial controversy over the nature of A_1 and A_2 specificity. For many years it has been known that antisera specific for A_1 red cells could be obtained by absorbing anti-A antisera with A_2 erythrocytes. This technique is used widely in blood grouping. Kisalius and Kabat (1978) presented evidence that the Type I A chain is the determinant of A_1 specificity, and the Type II A chain is involved in A_2 specificity. Thus the A_2 specificity is determined by the Type II chain in which the A trisaccharide:

$$\begin{array}{c} \text{L-Fuc}\alpha\ 1 \\ \downarrow \\ 2 \\ \text{D-GalNAc}\alpha 1 \rightarrow 3\text{D-Gal} \end{array}$$

is linked to $\beta 1 \rightarrow 4$ to DGlc NAc, whereas the A_1 specificity is determined by the Type I chain in which this trisaccharide is linked $\beta 1 \rightarrow 3$ to DGlc NAc.

Other investigators (Watkins, 1980) claim that both chains can be present in A_2 substances and that the difference is essentially quantitative. This controversy should be amenable to resolution by isolating monoclonal antibodies with A_1 specificity and studies using synthetic antigens prepared with Type IA and Type IIA determinants for production and absorption of antisera.

The Gal$\beta 1 \rightarrow 4$GlcNac$\beta 1 \rightarrow 6$Gal . . . grouping has been shown to be one of the I blood group determinants. Absence of the H blood group determinant on this grouping means that the ABO blood group system does not interfere with the phenotypic expression of the *I* gene, and that the I antigen may be able to coexist with A, B, or H antigen on the same branched heterosaccharide core.

The idea that the secretor status of an individual depends on the expression of the *H* gene also derives support from experiments. Milk contains L-fucosyl—$\alpha(1 \rightarrow 2)$—D-galactosyl—$\beta(1 \rightarrow 4)$—D-glucose only in secretors (Grollman and Ginsberg, 1967), and an enzyme that transfers L-fucose in $\alpha(1 \rightarrow 2)$ linkage to β-galactosyl residues was found in secretors, but not in nonsecretors (Shen, Grollman, and Ginsburg, 1968). The Le-specified L-fucosyl transferase was also demonstrated in milk from women of red cell phenotype Le(a+b—) and shown to be absent from the milk of Le(a—b—) women (Grollman, Kobata, and Ginsburg, 1968). Moreover, the *N*-acetylgalactosaminyl transferase associated with blood group A character occurs in milk and submaxillary gland homogenates from both secretors and nonsecretors (Hearn *et al.*, 1968; Kobata *et al.*, 1968), indicating that failure to secrete A substance depends on the absence of the requisite H-active structure, and not on lack of expression of the A gene. If the hypothesis that the Bombay phenotype results from the absence of the *H* gene is correct, that is, the persons are homozygous *hh*, then the lack of H, and hence of A and B antigens on the red cells is explicable by the absence of the fucosyl transferase specified by the *H* genes at the site of synthesis of the red cell substances.

Much of the recent knowledge of these enzymes has come from studies of various secretions, especially milk, in which they occur in soluble form; further, milk oligosaccharides have been useful in assaying glycosyl transferases by serving as acceptors. Much of this work has been published by Ginsburg's group. Although the substances used in these studies were obtained from secretions and not red cells, the A and B enzymes found in the gastric mucosa and submaxillary glands of humans and baboons have been used to transform type O red cells to type A and B red cells, respectively, proving that the specificity of these enzymes is the same for the oligosaccharides on red cells as those present in milk.

Lemieux *et al.* (1981) have recently developed an ingenious computerized system of mapping blood group oligosaccharides. The method employs the artificial synthesis of the blood group determinant, or hapten, with subsequent inhibition of binding to monoclonal antibodies. Haptens that "fit" the antibody most avidly can then be coupled to proteins and used as immunogens to raise blood group antisera, or attached to an immunoabsorbent column for purification. They have prepared very high-titer antisera to the A, B, and Le determinants using artificial antigens. They chemically synthesized the oligosaccharide hapten of the desired blood type, coupled it to bovine serum albumin (BSA), and injected this into rabbits. Following appropriate immunoadsorption, the refined antibodies were highly effective in immunofluorescent localization of these antigens in pertinent tissue. Using chemically synthesized oligosaccharides, Kabat's group demonstrated that the antibody-combining site of the monoclonal anti-I Ma antibody includes the β-LacNac—OCH_2 CHO— portions of such structures as β-LacNac ($1\rightarrow6$) β-D-Gal or β-LacNac ($1\rightarrow6$) β-D-Gal or β-LacNac ($1\rightarrow6$) β-D-GalNac. This demonstrates a considerable refinement in deciphering the topography of a particular epitope.

In summary the *O*, *le*, and *h* genes have not been shown to produce functional glycosyl transferases; the *se* gene (or absence of *Se*) prevents the expression of the *H* gene in certain organs. The A and B gene products are glycosyl transferases that accept galactose and N-acetylgalactosamine, respectively, from an activated substrate such as uridine diphosphate-D-galactose (UDP-gal) and transfer it to the end of the chain of sugars. Unfortunately the enzyme products (i.e., glycosyl transferases) have not as yet been subjected to amino acid sequence analyses, and so the biochemical basis for the difference in their specificity is unknown. It is conceivable that they differ from one another by only a single amino acid.

It appears that the A_m and B_m phenotypes are caused by a reduced activity of the *A*- and *B*-gene specified glycosyl transferases in erythrocyte membranes. Normal levels of the enzymes are present in the saliva of these individuals; whether levels are normal in their serum is controversial.

The other blood group antigen systems are listed in Table 6.6.

M, N AND Ss BLOOD GROUPS

The MN blood groups were first described in 1927 by Landsteiner and Levine. They injected rabbits with different samples of human red cells and absorbed the resulting rabbit immune serum with other red cells. This procedure resulted in antisera that discriminated between two groups of individuals, one termed M+ (whose cells react with the antiserum) and the other M—. Subsequent investigations with other antisera disclosed the genetic allele of *M*, namely, *N*. Hence three genotypes exist: *MM*, *MN*, and *NN*, and three corresponding phenotypes, M, MN, and N. Subsequent investigations of human sera containing antibodies to some human red cells but not others defined two more blood group factors, *S* and *s*, which seem to represent genetic alleles and are closely linked to *M* and *N*. The genetic region involved has been assigned tentatively to the long arm of chromosome 2 on the basis of band deletion in a person typed as hemizygous (M/—). Many uncommon alleles or subloci associated with this system now indicate the existence of at least 27 antigenic variants. Table 6.7 lists the common genotypes and phenotypes of the MNSs system.

Knowledge of this system is still accumulating. Somewhat less than 1% of Negroes lack both S and s, and these S—s— individuals presumably have a gene producing a third blood group factor in this system whose antibody has not yet been detected. Most of the blood not agglutinable by anti-S or anti-s is also not agglutinated by anti-U, an antibody described by Weiner in 1963, which failed to agglutinate the cells of about 1% of New York Negroes. Although anti-U could most easily be visualized as a mixture of anti-S and anti-s, a few S—, s— cells are not U—. Another interesting antigen in this system is M^g, representing an allele of *M* and *N*, which is detectable by its corresponding antibody and is shown by genetic studies to be linked to the MN system. It is thought by some but

Table 6.6 Blood group antigens in human beings

System	Antigens detected by	
	positive reaction with specific antibody	positive reaction with one antibody, negative with another
MNSs	M, N, S, s, U, M^g, M_1, Tm, M^k, M^v, Hu, He, Mi^a, Vw(Gr), Mur, Hil, Vr, Ri^a, St^a, Mt^a, Cl^a, Sul, Sj	M_2, N_2, M^c, M^a, M^v, S_2
P	P_1, p^k *Luke	P_2
Rh	D, C, c, C^w, C^x, E, e, e^s(VS), E^w, G, ce, ce^s(V), Ce, CE, cE, D^w, E^T *LW	D^u, C^u, E^u, many other variant forms of D, C, c and e
Lutheran	Lu^a, Lu^b, Lu^{ab}	
Kell	K, k, Kp^a, Kp^b, Js^a, Js^b, Kp^o, K^w, KL, Ul^a	
Lewis	Le^a, Le^b, Le^x, Le^c	
Duffy	Fy^a, Fy^b, Fy^3	
Kidd	Jk^a, Jk^b	
Diego	Di^a, Di^b	
Yt (Cartwright)	Yt^a, Yt^b	
I	I, i_1, i_2, I^T, I^D	
Xg	Xg^a	
Burrell	Sm, Bu^a	
Dombrock Do	Do^a	
Vel	Vel_1, Vel_2	
Very frequent antigens (97%)	Ge, Lan, Co^a, Gy^a, At^a, Ge, Sd^a, Yus, En^a, Rg^a, Salis, and many others, published and unpublished	
Very infrequent antigens	Levay, Wr^a, Be^a, By, Sw^a, Good, Stolzfus, Bi, Tr^a, Wb, Bp^a, Rd, Ls^a, Box, Tr^a, Wb, Rd, Or, Ht^a, Gf^a, Wu, Balkan, Bell, Dahl, Hildebrand, Mancus, Swann, At^a, Dp, El, Gn^a, H_T, In^a, Kn^a, Owens, Sp_1, Vennena, Becker, Bx^a, By, Burrett, Chr^a, Hartley, Jn^a, Job, Sf, Simpson, Zd, Skjelbred, Ta^a, Wb; many others, both published and unpublished.	
Other antigens	Au^a (Auberger), Sm and Bu^a ("allelic"), Bg, Cs^a, Chido, Sd^a, Ul^a, Go^a	

* A genetically independent part of the system.

Table 6.7 The MNSs genotype and phenotype frequencies in Caucasians, as defined by reaction to anti-M, anti-N, anti-S, and anti-s

anti- M	N	S	s	Genotypes		Phenotypes	
+	−	+	−	*MS/MS*	0.0611	MS	0.2011
+	−	+	+	*MS/Ms*	0.1400		
+	−	−	+	*Ms/Ms*	0.0801	Ms	0.0801
+	+	+	−	*MS/NS*	0.0397	MNS	0.2777
+	+	+	+	*MS/Ns*	0.1926		
+	+	+	+	*Ms/NS*	0.0454		
+	+	−	+	*Ms/Ns*	0.2205	MNs	0.2205
−	+	+	−	*NS/NS*	0.0064	NS	0.0689
−	+	+	+	*NS/Ns*	0.0625		
−	+	−	+	*Ns/Ns*	0.1517	Ns	0.1517

not proved that the M^g gene results in production of tiny amounts of M and a slightly greater amount of N. Other alleles such as M^k and M^v, whose antigens are also detected by corresponding antisera, are genetically linked to the MNSs system; such cells do not react with conventional anti-M reagents. Recently it has been demonstrated that incompatibility associated with M or N antigens can cause an HLA-restricted proliferative response in the primed lymphocyte test (PLT). This might explain why many HLA-D identical individuals stimulate each other in the mixed lymphocyte culture (MLC).

Several other rare antigens are genetically related to the MNSs system, including Mt^a (Miltonberger), Vw (Verwyst), Mu (Murrell), and Hil (Hill). (These terms, incidentally, indicate the standard nomenclature of blood group antigens; new antigenic determinants are usually named after the donor whose serum contained the first example of the antibody defining the antigen.) Many of the antibodies against the infrequent antigens of the MNSs system are common. For example, in 1957 nine examples of anti-Vw were found in 402 sera; at least five, and probably all, were naturally occurring. Anti-M^g occurs fairly commonly in human sera. Anti-Mt^a and anti-M^g have also been produced in rabbits. Biochemical studies designed to elucidate the chemical basis of the genetic differences in the MN and Ss antigens are currently in progress in several laboratories; preliminary data indicate that, like the ABO (H) and Lewis antigens, very complex carbohydrate moieties are involved in the antigenic determinants. (The M and N blood group substances appear to be sialoglycopeptides.) It appears that the immunodominant receptors of the MN system are developed by the addition of a "basal substance" of neuraminyl groups by the different neuraminyl transferases, one giving rise to the N antigen (detected by the lectin *Vicia graminea*), the other to the M antigen (reactive with *Iberis amera*).

The MN substances are glycoproteins with no significant amount of chloroform–methanol-extractable lipid (Springer *et al.*, 1966). The antigens are also myxovirus receptors (Springer and Ansell, 1958). Molecular weights from 1.5×10^5 to 12×10^6 have been found for these substances. MN antigens were originally thought to occur in nature only as substances with a molecular weight of 30,000; each of these molecules was thought to be held only via hydrophobic bonds in a lipidic layer of the erythrocyte membrane. However, the molecular weight of highly purified blood group MN antigens from human erythrocytes has been shown to depend on the method of isolation and the nature of the solvent. The molecular weight of the active repeating subunits isolated by gentle procedures has been shown to be 50,000 and 53,000 when the antigens were dissolved in 9 M formamide and 5% ethanol–water, respectively. Isolation at high pH or temperatures above 50°C yields preparations of molecular weight 30,000; these latter, smaller, only weakly active molecules may be degradation or disruption products of strong bound dimers, with both hydrogen and hydrophobic bonding helping to stabilize the aggregate (Kirschbaum and Springer, 1975).

More recent data based on cyanogen bromide cleavage and subsequent amino acid analysis indicate that the MNSs system may determine two structurally similar proteins, glycophorin A and glycophorin B. Springer *et al.* (1977) suggest that the *M* and *N* alleles apparently specify a difference in amino acid sequence at two positions of glycophorin A. The *M* allele substitutes serine and glycine at positions 1 and 5, respectively; the *N* allele places leucine and glutamic acid at these positions. (Earlier data suggested that oligosaccharides were the epitopes of M and N; thus M contained sialic acid and N contained galactose plus sialic acid. We suggest that this results from conformational interactions between these sugars and the underlying polypeptide backbone.) The allelic S and s epitopes appear to be carried on glycophorin B. This sialoglycoprotein shares the first 23 residues with A^N regardless of the MNSs type of the red cell from which the protein was isolated. Absence of either glycophorin A or B is not associated with any functional abnormalities, although no individual lacking both proteins has been discovered.

Of additional interest is the finding that all humans possess antibodies to a carcinoma-associated antigen that is a precursor of the MN antigens. For further details of the immunochemistry of the very complex system the reader is referred to Pardoe *et al.* (1971).

THE P BLOOD GROUP SYSTEM

The P blood group was also discovered by Landsteiner and Levine in 1927. They injected human cells into rabbits; the antibody so produced agglutinated the red cells of approximately 75% of the population. The positive cells were termed P positive, the remainder being P negative. In 1955 Sanger showed that a rare antibody, anti-Jay (Tj^a), which reacted with almost all cells tested, was in fact part of the P system. All of the rare Tj(a−) red cells were found to be P negative, and the antibody, when absorbed with P-negative cells, now reacted as anti-P. Vos *et al.* (1961) have found a hemolysin that resembles the classical anti-Tj^a in the sera of pregnant women who threatened to abort in Western Australia. This antibody resembles the autoimmune hemolysin found in paroxysmal cold hemoglobinuria (PCH). The true antibody 7S nature of this substance has not yet been determined. The nomenclature was modified, bringing anti-Tj^a, now known as anti-P+P_1, into the system. Three phenotypes were postulated: P_3, P_2, and p (see Table 6.8).

Table 6.8 The P blood group system

Phenotype	Frequency	Antigens on erythrocytes	Antibodies in serum
P_1	75%	P_1, P	none
P_2	25%	P	anti-P_1
p	very rare	none	anti-P_1PP^k*
P_1^k	very rare	P_1, P^k	anti-P
P_2^k	very rare	P^k	anti-P

* These sera agglutinate erythrocytes of all phenotypes except p, but they do not necessarily contain three separable monospecific antibodies; that is, specific anti-P^k reagents can be made by absorption of some sera with P_1 erythrocytes, but anti-P^k and anti-P_1 antibodies are removed from other sera by this procedure.

Nearly all P_2 individuals have anti-P_1 present in their serum as a naturally occurring antibody. An unusual variant of P that occurs in Finnish, and probably Japanese, people is termed P^k. These individuals have in their serum an antibody designated anti-P that reacts with P_1 and P_2 cells but is negative with p and other P^k cells, in contrast to anti-P+P_1 (Tj^a), which reacts with P^k cells as well as with P_1 and P_2 cells. The rarest form of autoimmune hemolytic anemia, paroxysmal cold hemoglobinuria, has been shown to be usually of auto-anti-P specificity.

Hydatid cyst (*Echinococcus granulosus*) fluid containing live scoleces will inhibit anti-P_1. Once lyophilized and stored, this product serves as a useful reagent in confirming the presence of anti-P_1 by inhibition tests.

Marcus and co-workers (1981) isolated a glycosphingolipid with P_1 activity from human erythrocyte stroma; the proposed structure, a chain of five sugars based on the structure of paragloboside with another sugar added, is given in Table 6.9. Watkins and her group (1981) also recently determined the structure of the terminal trisaccharide of the P_1 determinant of a glycoprotein isolated from hydatid cyst fluid, and that trisaccharide is identical to the terminal trisaccharide of the glycolipid.

Naiki and Marcus (1975) have also identified the P^k antigen as ceramide trihexoside (CTH) and the P_2 antigen as normal brain tissue globoside, on the basis of hemagglutination inhibition data. These two glycolipids are among the more abundant neutral glyco-

Table 6.9 Structures of glycosphingolipids*

Ceramide dihexoside (CDH)	Gal(β,1→4)Glc-Cer
Ceramide trihexoside (CTH (P^k)	Gal(α,1→4)Gal(β,1→4)Glc-Cer
Globoside—human red cells (P)	GalNAc(β,1→3)Gal(α,1→4)Gal(β,1→4)Glc-Cer
Globoside—rat kidney	GalNAc(β,1→3)Gal(α,1→3)Gal(β,1→4)Glc-Cer
P_1 antigen	Gal(α,1→4)Gal(β,1→4)GlcNAc(β,1→3)Gal(β,1→4)Glc-Cer
Paragloboside	Gal(β,1→4)GlcNAc(β,1→3)Gal(β,1→4)Glc-Cer
Ganglioside	NAN(α,2→3)Gal(β,1→4)GlcNAc(β,1→3)Gal(β,1→4)Glc-Cer
Forssman	GalNAc(α,1→3)GalNAc(β,1→3)Gal(α,1→4)Gal(β,1→3)Glc-Cer
Asialo G_{M2}	GalNAc(β,1→4)Gal(β,1→4)Glc-Cer

* Abbreviations: Gal = D-Galactose; Glc = D-Glucose; GalNAc = N-acetyl-D-galactosamine; GlcNAc = N-acetyl-D-Glucosamine; NAN = N-acetylneuraminic acid; Ceramide = N-acylsphingosine.

lipids of the erythrocyte (and other cells), and their structures have been established for some years. These data suggest that P^k erythrocytes, which lack the P antigen (Table 6.9), should lack globoside, and the p erythrocytes should be deficient in globoside and CTH. This prediction was confirmed by analysis of neutral glycolipids from two samples of P^k and two samples of p erythrocytes (one of each from the United States and from Japan). The P^k cells have no globoside and greatly increased CTH, and p cells have no detectable CTH or globoside. These are the only human red cells known to lack these glycolipids; the cells appear to be normal structurally, and the patients do not have hemolytic anemia, and so forth.

With regard to genetics and biosynthesis, no direct data on the enzymes yet exist, but it seems obvious that the products of the P_2 gene should be a β-*N*-acetylgalactosaminyl transferase that converts CTH (P^k) to globoside (P). The P^k and P_1 determinants have identical terminal disaccharides, and one can explain their synthesis two different ways: (1) a single α-galactosyl transferase that adds the terminal sugar to both determinants or (2) two different transferases (Fig. 6.7). The one-transferase theory readily explains the p phenotype with a single genetic defect—α-galactosyl transferase—but requires a more elaborate explanation for the normal P_1 and P_2 phenotypes. One reasonable hypothesis is that the P^k gene codes for an α-galactosyl transferase that synthesizes P^k (CTH) but does not effectively synthesize P_1 from paragloboside. The P_1 gene produces a regulatory molecule (enzyme subunit) that alters the acceptor specificity of the transferase and enables it to synthesize P_1 as well as P^k. The two-transferase hypothesis readily explains the P_1 and P_2 phenotypes, but to explain the p phenotype, the P^k and P_1 genes must be closely linked so that they are both affected by a single genetic lesion. The P^k phenotype results from a deficiency in β-*N*-acetylgalactosaminyl transferase activity, leading to an increase in CTH and lack of globoside.

The antigen p is considered by Naiki and Marcus to be the precursor of P_1 if the terminal sialic acid is removed with sialidase and another galactose is added to form the fifth sugar. However, P. Levine (personal communication) has questioned this view, since it would require a far higher frequency of antigen p than 1 : 150,000 for the genotype *pp*. It is simpler to consider paragloboside as a precursor to P_1, since this change can be effected by a one-step addition of a galactose, which is analogous to the O→A or B→A mutation in malignancy.

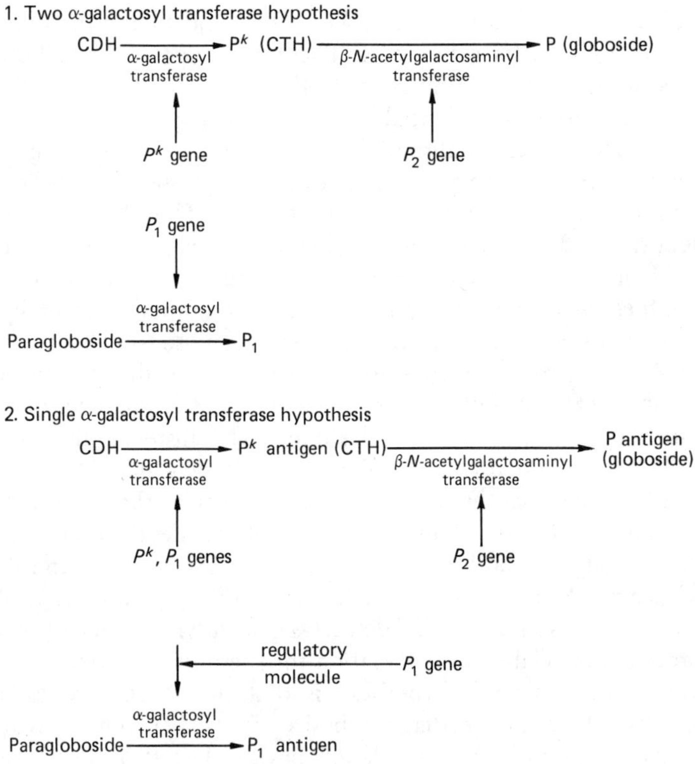

Fig. 6.7 Theories to explain the synthesis of P^k and P_1 blood group determinants.

THE Rh BLOOD GROUPS

Rh differs considerably from the ABO, M, N, Ss, and P blood group systems. First, its chemical structure has not as yet been determined. Second, it is not associated with naturally occurring serum antibody and thus cannot be evaluated by independent testing of serum. Finally, there are fewer Rh sites on red cells than sites for A or B, and as a result a different test system is needed to determine the presence or absence of Rh antigen on the red cells and of Rh antibody in the serum of persons who may have been immunized.

Rh antibodies occur in humans only after immunization with red blood cells of nonself genotype. This may result from exposure to maternal blood during pregnancy or after blood transfusion. The immune response is almost always IgG, and binding of about 20,000

IgG antibody molecules per red cell is required before agglutination can be observed. Since the average Rh-positive red cell has only 1 to 4×10^6 available sites for binding, Rh-positive erythrocytes suspended in physiological saline (0.85% of NaCl) cannot be effectively cross-linked by specific IgG antibodies as can type A or B red cells.

The Rh blood groups are extremely complex and have been the subject of considerable controversy in terms of both nomenclature and genetic mechanism. Fisher (1944) and Race and Sanger (1975) postulated three closely linked loci for the Rh antigens, and Wiener (1946) a single genetic locus. Accordingly, their nomenclature is different (Fisher using *C*, *D*, and *E* loci; Wiener, the *Rh* locus). The usual gene complexes and the antigenic mosaic produced by them are listed in Table 6.10, which includes an alternate terminology. In general most Rh antisera are "incomplete" and recognize one of the antigenic mosaics comprising each of the at least 27 different common (and 6 or more rare) Rh antigen complexes. Although the Fisher nomenclature is easier to comprehend, it assumes three closely

Table 6.10 The Rh gene complexes

Fisher	Wiener
CDe	R^1
cde	r
cDE	R^2
C^wDe	R^{1w}
cdE	r''
Cde	r'
CDE	R^z
C^wde*	r'^w
CdE	r^y
C^wDE*	R^{zw}
C^wdE*	r^{yw}

The antigens determined by these gene complexes are inherited in fixed combinations, that is, phenogroups. Thus, a mating of an R^1R^1 parent with an rr parent produces offspring all of whom are R^1r. Both the R^1r (*CDe/cde*) and an R^or^1 genotype will result in identical phenotypes, that is, cells that react with anti-C, anti-D, anti-e, and anti-c. Similarly a person whose cells react with the other four antisera plus anti-E may be R^2r (*cDE/cde*), or R^1R^2 (*CDe/cDE*). Genotypes are expressed by superscripts.

* Habibī and co-workers (*Vox Sang.* 31:103, 1976) have recently reported evidence suggesting that the C^w antigen is in fact heterogeneous.

Table 6.11 Rh nomenclature equivalents

Rh1	Rh_o, D	Rh12	rh^G, G	Rh23	D^w (Wiel)
Rh2	rh′, C	Rh13	Rh^A	Rh24	E^T
Rh3	rh″, E	Rh14	Rh^B	Rh25	LW
Rh4	hr′, c	Rh15	Rh^C	Rh26	"c-like"
Rh5	hr″, e	Rh16	Rh^D	Rh27	cE
Rh6	hr, f, ce	Rh17	Hr_o	Rh28	hr^H
Rh7	rh_i, Ce	Rh18	Hr	Rh29	RH, "Total Rh"
Rh8	rh^{w1}, C^w*	Rh19	hr^S	Rh30	Go^a, D^{Cor}
Rh9	rh^x, C^x	Rh20	VS, e^s	Rh31	hr^B
Rh10	hr^V, V	Rh21	C^G	Rh32	Troll (Reynolds)
Rh11	rh^{w2}, E^w	Rh22	CE	Rh33	R_o^{Har}

* Habibi and co-workers (*Vox Sang.* 31:103, 1976) have recently reported evidence suggesting that the C^w antigen is in fact heterogeneous.

linked loci, whereas the Wiener nomenclature assumes only one locus with multiple alleles. Evidence favoring the Wiener hypothesis includes (1) failure to find anti-d, despite a diligent search for 40 years, and (2) the evidence of Toma *et al.* (1979) using uranyl-labeled anti-Rho and ^{99m}technetium-labeled anti-Rho in immunoelectron microscopy that rr(dd) cells contain about 10% as many antibody-binding sites for anti-Rho (anti-D) as do Rho-positive cells; they suggest that the lack of detectable agglutination of these cells is a quantitative rather than a qualitative phenomenon.

The individual antigens of which the mosaics are composed are recognized by sera from humans immunized either deliberately or as a result of an Rh-incompatible pregnancy or transfusion; these antigens are listed in Table 6.11 under three terminologies. Anti-d antiserum does not exist, and the "presence" of d is inferred by negative reactions to anti-D, that is, to anti-Rh_o (there probably is no d antigen). Variants of each of the components of the Rh mosaics have been recognized by appropriate antisera; the most important of these is the antigen D^u, which is a weak D. There are many varieties of D^u, which is especially common in Negroes. D^u is important because it can stimulate the formation of anti-D (that is, anti-Rh_o) in an Rh_o-negative person and also because some apparently D^u individuals have made anti-Rh_o. [The modern view is that these people are in reality lacking a specific component of the (Rh_o) antigen mosaic.] Compound antigens also exist, for example, for ce; that is, an anti-f(ce) antibody reacts with cells containing both c and e when *c* and *e* are present on the same chromosome (*cis* position), but not when on separate chromosomes (*trans* position).

Most anti-Rh_o antibodies are "incomplete," but a few produce direct agglutination. Direct agglutination is almost always due to IgM anti-Rh_o, whereas the "incomplete" antibodies are almost always IgG (a few IgA anti-Rh_o antibodies have been described). Rh_o cells strongly coated with incomplete anti-Rh_o block agglutination of the cells by "complete" IgM anti-Rh_o. Hence the "incomplete" antibodies are sometimes termed "blocking" antibodies. ("Incomplete" anti-Rh antibodies were formerly thought to be univalent, that is, to possess only one antigen-binding site. This theory has been disproved.)

In an ultrastructural study using ferritin-labeled anti-Rh in albumin and papain tests, Voak *et al.* (1974) found that strong hemagglutination showed large areas of cell-to-cell contact, whereas neuraminidase caused only loose agglutination with small areas of cell-to-cell contact, confirming a previous suggestion that reduced zeta potential is not the significant factor in Rh hemagglutination. The ferritin-labeled anti-D was clustered only on neuraminidase- and papain-treated cells; thus these enzymes may cause limited membrane mobility. The authors postulated that the loose agglutination in neuraminidase tests was due to localized high density of D sites and that the much greater agglutinating activity in papain tests was due to clustering and protease activity. Increased anti-D antibody uptake was not a critical factor in enzyme tests, because cells sensitized and then washed before enzyme treatment were strongly agglutinated. Other studies by Voak *et al.* (1971) revealed that dextran does not reduce zeta potential but acts by adsorption between adjacent cells, and the authors suggested that a similar mechanism might explain the action of albumin.

Particularly important are rare bloods lacking certain of the antigens of the Rh mosaic, for example, D bloods. Such individuals are usually found to be offspring of consanguinous marriage. Rh_{null} blood (that is, — — — cells) are also of considerable interest. The Rh_{null} bloods, which lack any Rh antigen, derive from two different genetic backgrounds. One is a homozygote of the amorphic Rh allele *r*(— — —), designated type rh (double bar Rh) (— — —/— — —, Rh:—29), whose sole example is the Japanese propositus (H.A.) studied by Hasekura *et al.* (1971). The other is a homozygote of the suppressing gene $X^{o}r$ independent of the *Rh* gene loci, which inhibits the whole Rh antigens and the U antigen of the MNSs system. This suppressed type of Rh_{null} is designated rh (zero Rh) or rh_m (Rh:—29) and is found among various populations.

Because these suppressed Rh_{null} bloods are found to be associated with hemolytic anemia, attention has been focused on the question whether rh (or Rh: —29) is also hemolytic. The rh(— — —/ — — —) subject has slightly increased hemolysis without anemia associated with red cell abnormalities (stomatocytosis and spherocytosis). Antibodies to red cells found in warm antibody hemolytic anemia can be defined on the basis of their reaction or lack thereof with Rh_{null} cells, cells with partially deleted Rh antigens, or cells containing normal Rh antigens.

The Rh system is also of great clinical importance, since the Rh_o(D) antigen seems to be the most immunogenic of all the various blood group factors. Hence hemolytic disease of the newborn is most frequently due to maternal antibodies to Rh_o when an infant whose red cells contain this antigen (due to a gene inherited from the father) is born to an Rh_o-negative mother. However, in the absence of prior sensitization of the mother by incompatible transfusion, firstborn infants are rarely affected. Presumably mothers are sensitized at delivery of the first Rh-incompatible infant; during subsequent pregnancies anti-Rh_o titers, detectable by standard serological means, appear at about 3 months' gestation and gradually increase. Work by several groups shows that the initial "sensitization" (e.g., immunization) can be prevented by administering exceedingly small quantities of anti-Rh_o during the first pregnancy of Rh-negative females mated to Rh-positive males. These findings are important in two respects: (1) they shed great light on feedback mechanisms governing shutoff of antibody synthesis, and (2) their clinical application is gradually resulting in the elimination of hemolytic disease of the newborn due to anti-Rh incompatibility. Little is known of the chemistry of the Rh antigens, but they appear to be phospholipids. Degradation of red cell stroma phospholipids by phospholipase A_2 is accompanied by a fall in the activity of the Rh antigens c, D, and e; indeed Hughes-Jones *et al.* (1975) have shown that not only phospholipase A_2 but also phospholipase C, which converts phospholipids to diglycerides, can inactivate the D antigen. Anti-D bound to the red cells protects the D antigen from inactivation by phospholipase A_2.

The number of available antigen sites with specificities D(Rh_o), C(hr′), e(hr″), and E(rh″) have previously been estimated (see recommended article by Hughes-Jones *et al.*), and by the use of trace-labeled antibody the number of available G antigen sites has recently been estimated (see Skov, 1976). The G antigen is detected in optimal amounts on C-positive samples; when D is present without C, the number of G sites is apparently dependent on the expression of all the D antigens within the recognized six categories of D.

LW

Landsteiner and Weiner (1940) injected rhesus monkey red cells into guinea pigs and found that the resulting antisera, after partial absorption with human Rh-negative red cells, would react strongly with (adult) Rh-positive red cells and weakly with (adult) Rh-negative red cells. Although the antibody produced seemed to be identical with human anti-Rh, we now know that it was directed against a different antigen specificity, termed LW. Evidence that anti-LW is different from anti-Rh was obtained by Fisk and Foord (1942) when they showed that anti-Rh sera react equally strongly with Rh-negative and Rh-positive red cells from umbilical cord bloods. Further they found that extracts from either Rh-negative or Rh-positive red cells would stimulate the formation of "anti-Rh" in guinea pigs.

Anti-LW can be eluted from D-negative cells, even though these are not agglutinated. Levine *et al.* (1963) produced further evidence of the difference between anti-D and anti-LW when they found a D-positive, LW-negative woman with anti-LW in her serum; she presumably had been immunized by previous pregnancies and transfusions. Her red cells were not agglutinated by guinea pig anti-LW, described above; their injection into guinea pigs failed to elicit the formation of anti-LW.

The gene responsible for LW is inherited independently of the Rh genes, although evidently there is a close relationship between the LW and the Rh gene products. This was illustrated by the observations of Levine *et al.* (1962) that red cells of type Rh_{null}, injected into guinea pigs, failed to produce "anti-D-like" substances (e.g., anti-LW) in 20 of 20 cases, although the injection of ordinary Rh-negative cells (LW-positive) produced antibody in 12 of 20 cases.

The relationship between LW and Rh is still unproved, but Race (1968) suggested the scheme illustrated in Figure 6.8 to explain the relation between the *LW* and *CDE* genes: this is the current working hypothesis of some people in the field but probably is not valid.

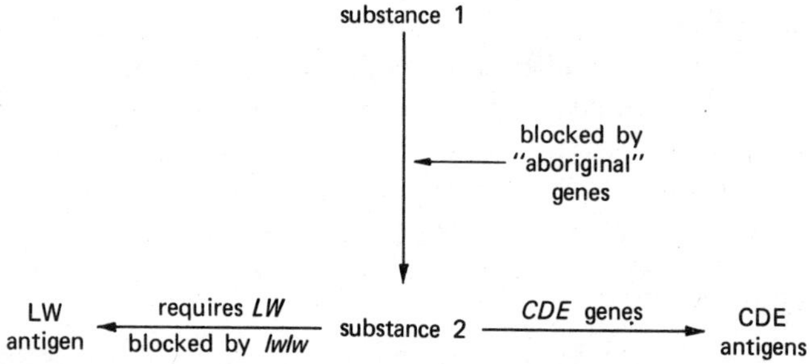

Fig. 6.8 A tentative scheme for the genetic pathways leading to the antigens of the Rh complex. (After R. Race and R. Sanger, *Blood Groups in Man*, 5th ed. Philadelphia: F. A. Davis Co., 1968, p. 217. By permission of the author and publisher.)

Inheritance of a double dose of an autosomal recessive gene (termed $X^{o}{}_{r}$ by Levine) prevents formation of a second precursor substance, producing LW-R_{null} cells. Two separate genes are involved in formation of the LW and Rh antigen from the first precursor substance; that is, homozygosity for lw results in LW-negative cells with normal Rh determinants. Incidentally, the *Rh* gene has been mapped on chromosome 1, as has the Duffy blood group gene locus.

THE KELL SYSTEM

The Kell system when first described appeared to be a very simple two-antigen system. In 1946 Coombs *et al.* discovered anti-Kell; indeed it was the first new blood group factor to be disclosed by the new antiglobulin test. In 1949 the antithetical antibody anti-k (originally called anti-Cellano) was described by Levine *et al.* The red cells of 9% of the English population were found to react with anti-K (that is, were K+), having the genotype *KK* or *Kk*; only 0.2% of these were *KK*; 99.8% of the population reacted with anti-k, 91% being *kk*. The simple concept of a two-antigen Kell system became obsolete in 1958 when Allen, Lewis, and Fudenberg, working with the serum of a Mrs. Penney (anti-Kp^a) and later the serum of a Mr. Rautenberg (anti-Kp^b), made two observations that led to Kp^a and Kp^b being added to the Kell system. The anti-Kp^a serum reacted only with 2% of the population; however, of the first 73 Kp+ individuals found, only 2 were K+. If Kp^a were not related to Kell, 7 of the 73 (i.e., 9%) of the bloods sampled should have been K+. The second observation was that in three unrelated families all persons positive for both Kp^a and K were k-negative, suggesting that in these people the paired loci were occupied by *K* and Kp^a instead of *k*. The evidence suggested that Kp^a was allelic to *K*. The antithetical antigen to Kp^a recognized by anti-Kp^b is present in more than 99.9% of the population. In 1957 Chown *et al.* described a new phenotype in the Kell system, K—k—Kp(a—b—), known as K_o. The serum of this propositus (Celtz) contained an antibody that reacted with all K+ or k+ samples. The antibody was called anti-Ku. It is assumed that the gene responsible for the K_o phenotype is an allele at the Kell locus because several families have been found that can be interpreted only by postulating an allele in a heterozygous state, producing no K, k, Kp^a, or Kp^b antigen (Fig. 6.9). In 1965 Stroup *et al.*, while studying a new antibody anti-Js^b (Matthews) antithetical to Js^a (Sutter), noted that all K_o cells tested were also Js(a—b—). The probability that this could happen by chance alone was 10^{-15}.

It seems then that control of the antigens in the Kell system may be similar to that of the Rh system, in which antigens are determined by multiple alleles at a single locus or by sets of alleles at very closely linked loci. In the Kell system the pairs of alleles *K* and *k*, Kp^a and Kp^b, and Js^a and Js^b, seem to occupy three closely linked loci.

Some other blood samples have been described as possibly belonging to the Kell system: McLeod, Williams, Claas (KL), and Karhula (Ul^a).

Fig. 6.9 The Kell system: a scheme, ripe for toppling, that can, however, help in trying to remember the present complicated facts. (From R. Race and R. Sanger, *Blood Groups in Man*. 6th ed. Philadelphia: F. A. Davis Co., 1975. By permission of the authors and publisher.)

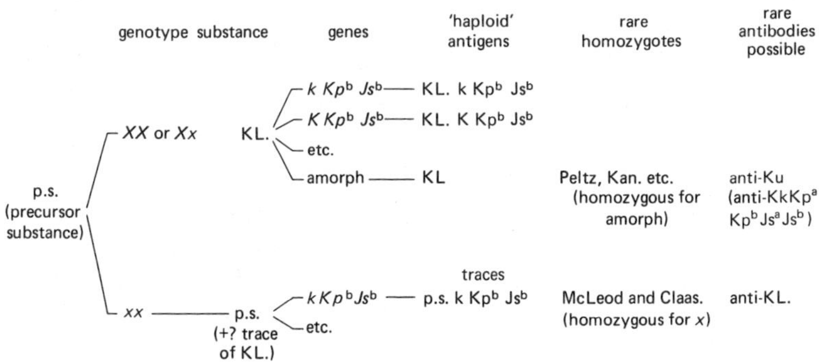

The story is not yet complete. Race and Sanger (1968) have proposed a scheme to fit some of these rarities into the Kell system. A possible derivation of the various antigens from a precursor substance is depicted in Figure 6.9. In 1961 Allen and Rosenfield proposed a numerical nomenclature for the Kell system (see Table 6.12), analogous to the numerical nomenclature for the Rh system. Recently Yamaguchi *et al.* (1979) detected a new Kell complex antigen, K_p^c, in four sisters who were the offspring of a consanguineous mating.

The Kell system is of clinical importance, since, after the ABO and $Rh_o(D)$ antigens, the K antigen probably has the most potent antigenicity. From 5 to 10% of K— recipients will form anti-Kell following a single transfusion of K+ blood (Giblett, 1961; Kornstead and Heisto, 1957). The antibody, normally best detected by the Coombs test, can cause hemolytic transfusion reactions and hemolytic disease of the newborn.

Of the 18 red cell antigens associated with the Kell blood group, one, named Kx, is the product of an X-linked gene and appears to be a precursor in the Kell biosynthetic pathway. Lack of Kx on red cells, caused by inheritance of a variant allele at the X-linked locus, results in gross changes in Kell antigenicity, an effect called the McLeod phenotype (Giblett *et al.*, 1971). Such cells also show striking morphological changes. Normal phagocytic leukocytes lack Kell antigens but have strong Kx. The leukocytes of boys with X-linked chronic granulomatous disease lack Kx antigen and have defective bactericidal function. The fundamental defect in chronic granulomatous disease appears to be failure to inherit the X-linked gene that determines Kx synthesis (Marsh, 1977a).

THE DUFFY BLOOD GROUP SYSTEM

The first Duffy blood group antigen was described in 1950. The antibody was called anti-Fy^a, and its antithetical partner, found in 1951, anti-Fy^b. With these two antisera four phenotypes are observed.

Phenotypes	*Genotypes*
Fy(a+b—)	Fy^aFy^a
Fy(a+b+)	Fy^aFy^b
Fy(a—b+)	Fy^bFy^b
Fy(a—b—)	?$FyFy$

Table 6.12 Nomenclatures and some gene complex products in the Kell system. *

		Antigens produced									
New name		K1	K2	K3	K4	K5	K6	K7	K8	K9	
Original name		K	k	Kp^a	Kp^b	Ku	Js^a	Js^b	K^w	KL	
First antibody		Kell	Cellano	Penney	Rautenberg	Peltz	Sutter	Matthews	Williams	Claas	
Gene complex											
Old name	New Name										
$K\ Kp^b\ Js^b$	K	+	o	o	+	+	o	+	o	+	
$K\ Kp^b\ Js^b\ K^w$	K^w	+	o	o	+	+	o	+	+	+	***1
$k\ Kp^b\ Js^b$	k	o	+	o	+	+	o	+	o	+	
$k\ Kp^b\ Js^b\ K^w$	k^w	o	+	o	+	+	o	+	+	+	
$k\ Kp^a\ Js^b$	k^p	o	+	+	o	+	o	+	o	+	
$k\ Kp^b\ Js^a$	k^s	o	+	o	+	+	+	o	o	+	
K_o	k^o	o	o	o	o	o	o	o	***2	+	
Phenotype not a gene complex ***3											
McLeod		o	+ weak	o	+ weak	***4	o	+ weak	o	o	

* So far, 18 distinct Kell antigens have been described. The most recent, K18, was detected in all but 1 of 2000 bloods tested, and the one exception was a blood of K_0 phenotype containing anti-K18 activity.

***1 It is not yet known whether any of the gene complexes that produce Kp^a or Js^a produce K^w. If such complexes exist they will be additional to the ones shown in this table.

***2 Of 5 K_o bloods tested, 4 were K^w-positive and one K^w-negative.

***3 There is no clear indication of the genetic background of the McLeod phenotype. Corcoran *et al*. thought that the condition might result from the presence of independently inherited "suppressor" or "inhibitor" genes. If this is so it seems that the genes involved are recessive.

***4 The original Peltz serum reacted with McLeod cells. The anti-Ku prepared by absorption and elution from the Peltz serum did not react. An antibody identified as anti-Ku by Nunn *et al*. did react with the McLeod cells. It seems that if the McLeod cells are Ku positive at all, their Ku expression, like their k, Kp^b, and Js^b expression, may be considerably weakened.

(This table is modified from one prepared by Dr. F. H. Allen, Jr., and is reproduced here with his permission.)

The Fy(a—b—) phenotype is rarely found in Caucasians but occurs in 68% of Negroes. It has been suggested that a third allele *Fy* exists, as well as Fy^a and Fy^b, that produces the Fy(a—b—) phenotype. Sometimes sera are encountered containing antibodies that react only with Fy(a—b—) cells. It has been suggested that at least two forms of the *Fy* gene exist. The common one in Negroes produces neither Fy^a nor Fy^b, but the other form, termed Fy^x, produces an antigen that reacts weakly with some examples of anti-Fy^b. The Duffy system is of great interest to anthropologists. Of all the known blood group genes *Fy* makes the greatest distinction between Negroes and whites—with a frequency of probably more than 90% in West Africans, 80% in New York Negroes, and about 3% in Europeans. The phenotype Fy(a—b—) is common in Yemenite Jews, whereas the phenotype Fy(a+) was found in all but 1 of 145 Japanese and all but 1 of 394 Koreans.

LUTHERAN BLOOD GROUP SYSTEM

The antibody defining the first Lutheran antigen Lu^a was discovered in 1945 in a multiply transfused patient. The antibody agglutinated red cells of 8% of the population. In 1956 its antithetical partner anti-Lu^b was found. With these two antisera four phenotypes were observed.

Phenotypes	*Genotypes*
Lu(a+b—)	Lu^aLu^a
Lu(a+b+)	Lu^aLu^b
Lu(a—b+)	Lu^bLu^b
Lu(a—b—)	

The rare Lu(a—b—) phenotype can arise from two genetic backgrounds. In one a gene with dominant effect prevents the expression of the Lutheran genes. In the other, individuals who are homozygous for a rare recessive gene have the Lu(a—b—) phenotype. The Lutheran locus and the secretor locus provided the first example of autosomal linkage in humans and, incidentally, may be the first example of autosomal crossing over. Recently, a new rare antigen Li^2 (Livesey) was found to be inherited as an autosomal dominant and appeared to be linked to the Lutheran system.

KIDD BLOOD GROUP SYSTEM

The first Kidd blood group antigen was discovered in 1951. The antibody was called anti-Jk^a and its antithetical partner, discovered in 1953, anti-Jk^b. These antisera define four phenotypes.

Phenotypes	*Genotypes*
Jk(a+b−)	Jk^aJk^a
Jk(a+b+)	Jk^aJk^b
Jk(a−b+)	Jk^bJk^b
Jk(a−b−)	?*JkJk*

The rare Jk(a—b—) phenotype has so far been described only in non-Caucasians. It has been suggested that a third and silent allele, *Jk*, may be responsible for this phenotype.

Xg BLOOD GROUP SYSTEM

In 1962 a new blood group antibody was found, anti-Xg^a. This antibody defined an antigen Xg^a, the gene for which was shown to be present on the X chromosome. This is thus far the only sex-linked blood group system. The phenotype frequencies for the two sexes are 88% Xg(a+) for females and 66% Xg(a+) for males. *Xg^a* is a dominant gene that obeys the rules of X-linked inheritance. Sons receive their X chromosome from the mother and the Y from the father. Daughters receive an X from each parent. The mating of an Xg(a+) father and an Xg(a—) mother produces Xg(a—) sons and Xg(a+) daughters. Xg^a-hemizygous fathers and homozygous mothers produce sons and daughters who are Xg(a+). When both parents are Xg(a—), the children are Xg(a—). Daughters who are Xg(a—) must have Xg(a—) fathers. The Xg blood group system has been used in studies on aneuploidy (incorrect number of sex chromosomes). An incorrect number of sex chromosomes can be received from either parent. The irregularity or accident in cell divisions can occur during spermatogenesis or oogenesis. The derivation of an abnormal X chromosome number, whether paternal or maternal, in many cases can be determined by Xg^a typing of the families.

THE Ii BLOOD GROUP SYSTEM

The vast majority of adult human red cells (greater than 99.99% of Caucasians) contain an antigen that reacts with the cold agglutinins in the serum of patients with high-titer cold-agglutinin hemolytic anemia. A few normal adult red cells are not agglutinated by such sera and are termed I-negative, or i cells. The antibodies of these hemolytic anemia patients are designated as possessing anti-I specificity. (Certain lectins can also be used to distinguish I-positive from I-negative cells, in that they react more strongly with I cells.) These anti-I antibodies react weakly with cells from umbilical cord blood, since at birth the I antigen is poorly developed. The reactions of anti-i are reciprocal to those of anti-I with the normal adult, the rare I-negative adult, and cord cells. It now appears that anti-i is quite common in the serum of patients with infectious mononucleosis. The relative ratio of I/i reactivity is altered in red cells of certain hematological conditions, for example, thalassemia. Further, cold agglutinins of certain aboriginal population isolates in New Guinea, Venezuela, and elsewhere give stronger reactions with cord cells (I^T) than with adult I or i cells; other aboriginal populations have cold agglutinins in high incidence that agglutinate I, i, and fetal cells. Cold agglutination of AI specificities (that is, reacting with A+I+, but not as well with OI or Ai cells) were found in high titers in from 48 to 94% of various population isolates in some Melanesian populations, and this was assumed to result from cross-reactions with bacterial or parasitic antigens.

During the first 18 months of life there occurs a gradual increase of red cell I antigen associated with a decrease of i, but adult red cells retain small amounts of i. Although the biochemical mechanisms involved are unknown, recent studies with secreted glycoproteins indicate that the determinants responsible for I and i specificities appear at intermediate stages of the biosynthesis of ABH and Lewis antigens. Quantitative precipitin studies with monoclonal anti-I and anti-i antibodies and a panel of blood group-related glycoproteins have indicated that there are at least six types of I and four types of i determinants. The i determinants have been shown to involve a nonreducing D-Gal—β(1→4)—D-GlcNac structure that constitutes the type II chain of blood group glycoproteins. The I-antigenic determinant is composed of various domains on the branched structure:

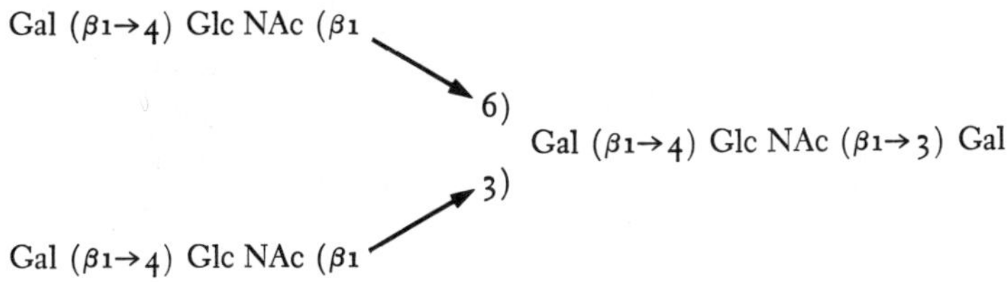

This structure is formed by the addition of Gal (β1→4) Glc NAc (β1→6) branch to the i-active structure.

The I/i antigens occur on the carbohydrate chains that constitute the core structures of the blood group ABH antigens and certain erythrocyte gangliosides. Generalizations on the molecular structure of I and i specificities may be summarized as follows: (a) Nonreducing terminal β-galactose is an important part of both the I- and i-antigenic determinants; (b) an intact straight-chain oligosaccharide with a repeating Gal β1→4 GlcNAcβ1→3 sequence expresses the majority of i and part of I determinants; (c) the antigenic determinants recognized by the majority of anti-I antibodies are expressed on intact 1→4 and prove to be widely distributed in nature, since the cow blood group substances also contain them. It should be emphasized that the heterogeneity of the water-soluble glycoproteins is quite extensive, probably because of incomplete biosynthesis or degradation.

By using certain degradation procedures (periodate oxidation and Smith degradation), substances with A and B reactivity have been converted to precursor substances with demonstrable I specificity. This supports the contention that I antigens are internal to the A, B, H, Le^a, and Le^b determinants. Likewise, I Ma determinants have been synthesized on the precursor obtained by Smith degradation of hog mucin substance (see Fig. 6.10).

Other cold agglutinin specificities exist, termed Pr_1, Pr_2, and so forth, based on the reaction of the antibodies with red cells treated with proteolytic enzymes.

Recently, both I and i have been found on macrophages, leukocytes, and some cultured cell lines by radioimmunoassay and immunofluorescence. I/i antigens were also detected on the epithelia of Henles' loops and in the distal tubules of human nephrons. The antigens reacting with the "cold-reacting" antibodies anti-I, anti-i, and anti-P (Donath-Landsteiner antibody) have been shown to be present on the glycophorin molecule (the major human erythrocyte membrane glycoprotein) by inhibition of antibody-induced lysis or

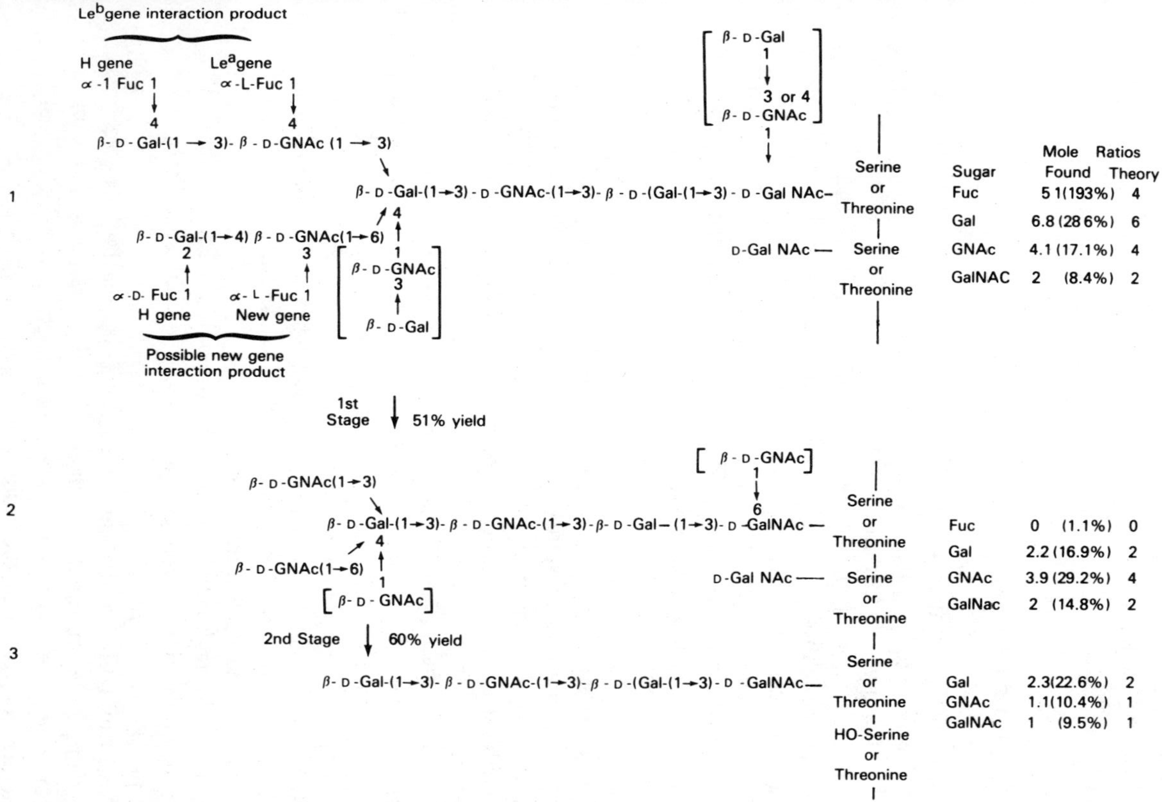

Fig. 6.10 Stepwise Smith degradation of ovarian cyst blood group H substance with periodate. (From E. A. Kabat, in *Blood Group Antigens,* D. Aminoff, Ed. New York: Academic Press, 1979, p. 187)

of antibody adsorption, by use of radiolabeled, purified antibodies. Although these antibodies do not react with red cells at 37°C, they react with isolated glycophorin at this temperature.

Some human cold agglutinins with $Pr_{1\text{-}3}$ and Gd specificities were found to react with glomerular capillaries and glomerular endothelium, respectively. These may contribute to hyperacute kidney transplant rejection in patients with high titers of cold agglutinins.

Other well-defined blood group systems are included in Table 6.6, but space limitations preclude discussion.

THE ANTIGENS DONNA, BENNETT, GOODSPEED (DBG), STURGEON, AND PRICE

These antigens are similar to each other but probably not identical. They are fairly uncommon, and antibodies that detect one of them usually detect several, but not always all, of the others. Seaman *et al.* (1967) divided anti-DBG antibodies into three specificities that they call anti-Bg^a, anti-Bg^b, and anti-Bg^c. In 15 of 530 Canadians (2.8%), some expression of the DBG appeared. With more sensitive methods at least part of the DBG antigen complex may be identified in as many as 30% of the general population, in whom most cells will be positive. Most cells in this 30% of people are positive for one or more, but not usually all, parts of the complex. Thus different examples of what have up to now been called anti-DBG react differently with cells that have been classified as DBG-positive. J. Tegoli, J. P. Harris, P. D. Issitt, and C. W. Sanders (unpublished observations, 1968) have recently encountered a bizarre antibody that reacts only with the cells of persons who have formed anti-DBG. If this antibody does, in fact, detect an antigen related in some way to the DBG complex, it is possible that this antibody and anti-DBG between them detect cells from persons homozygous for the antigen-controlling genes in the system. This would mean that heterozygotes (the majority of people) react with neither. Tegoli and co-workers did not give the antibody a name.

Inheritance of these antigens is not straightforward; a good example of the type of problem that arises is the blood Price, described by Dunsford (1962), where the propositus had an exceptionally well-developed antigen but whose parents both typed as negative.

Morton, Pickles, and Sutton (1969) have shown that the Bg^a and HLA-B7 leukocyte antigen are related. Anti-Bg^a appears to be directed primarily against the HLA-B7 antigens, which is a part of the major histocompatibility locus in humans. Absorption studies show that HLA-B7-positive leukocytes will totally absorb anti-Bg^a. It has not been possible to determine exactly the capability of Bg (a+) red cells in absorption of some anti-HLA-B7 sera. This has led Morton *et al.* to postulate that these antigens disappear as the red cells mature and that Bg^a may be that portion of the HLA-B7 antigen that survives red cell maturation.*

"PUBLIC" AND "PRIVATE" ANTIGENS

In addition to the well-defined blood groups, some antigens, the so-called "public" antigens, are almost ubiquitous in human populations (Table 6.6). When first found, a public antigen may indicate a new genetic system, or it may be part of a previously recognized system. U, for example, was not at first recognized as part of the MNSs system. Persons lacking these high-frequency antigens are presumably either (1) homozygous for the absence of a very common allele of a gene necessary for the production of the antigen (analogous to the Bombay gene in the ABO groups) or (2) amorphs unable to produce precursor substances. Among these very frequent antigens are Vel_1, and Vel_2, Ge, Lan, Co^a, At^a, and En(a). En(a) is especially interesting. The cells of the En(a—) antibody former and those of compatible En(a—) sibs are highly unusual, and it appears that the state of being En(a—) is associated with the inheritance of genes that cause some modification of the cell surface. En(a—) cells react more strongly with some and less strongly with other blood group antibodies. The En(a+) state (which includes the vast majority of "normal" blood types) seems to indicate that the genes responsible for the production of En^a are also responsible for the normal red cell membrane. The abnormality of the red cell surface of En(a—) cells has been likened to a probable abnormality of the red cell surface in the Rh_{null} condition.

In contrast some antigens are present so infrequently (incidence less than 0.25% in the populations studied) that they are called "pri-

* Although it seems probable that the anti-Bg^a antibodies are reacting with the antigens of the HLA-B7 group, there is insufficient evidence to decide if the HLA-B7 antigen is *identical* with the Bg^a antigen. So far all antisera containing anti-Bg^a hemagglutinating antibodies have also contained specific cytotoxic antibodies or such a complex mixture that it has been impossible to decide if anti-HLA-B7 is present. Family studies showed that the HLA groups may influence the Bg^a reaction of the red cells.

vate" antigens (see Table 6.6). Each is a dominant characteristic defined by a specific antibody, shown by serological means not to duplicate any others previously discovered. It is obvious that genes determining private antigens may have alleles that will determine antigens present in the same population in very high frequencies; for example, an antiserum defining Wr^b, the allele of the private antigen Wr^a, was discovered in 1971 (Adams *et al.*).

CLINICAL SIGNIFICANCE OF BLOOD GROUP ANTIGENS

Clinically blood group antigens are of most importance in blood transfusion (see Mollison, 1961, 1967). Using the enzyme α-galactosidase, type B erythrocytes have been enzymatically converted into type O cells, which were proved effective in transfusions to persons of all three blood types (O, A, and B). It is of interest that antibodies of known blood group specificity eluted from red cells of patients with acquired autoimmune hemolytic anemia fall into one of three blood group systems, that is, Rh, Kell, and MNSsU. To the author's knowledge maternal antibodies to antigens in these three systems are the only ones capable of causing death *in utero* of an infant due to hemolytic disease ("hydrops fetalis"). Vos *et al.* (1971) and Weiner and Vos (1963) found that antibodies against well-defined Rh antigens (CDEce) were infrequently found in the red cell eluates from patients with acquired hemolytic anemia of the warm type (AHA). In contrast, anti-nl (against all normal Rh red cells), anti-pdl (against partially deleted Rh red cells), and anti-dl (fully detected Rh red cells) were frequently present in red cell eluates. These findings suggest that the initial site of action of autoantibody development in AHA may result from a defect of the structural composition of the rhesus genome, which is thereafter rejected by a normal immune mechanism. The subsequent development of a more generalized autoimmune abnormality involving the formation of multiple autoantibody-forming cell lines does not necessarily indicate the establishment of an aberrant immune apparatus.

These antigens have also proved useful in linkage studies, particularly in chromosomal mapping. This is especially true of the Xg^a sex-linked antigen, which has been used in mapping the X chromosome by measuring recombination distances for various X-linked genes (e.g., in hemophilia and color blindness). There is good evidence that the Duffy locus (*Fy*) is on the first chromosome (which is also the locus for one form of hereditary cataract): 15 more genes are now firmly or tentatively assigned to chromosome 1, including the *Rh* locus. The *MNSs* locus appears to be on the second chromosome. Recent data indicate that the *ABO* genes seem to be linked to hereditary onycoosteodysplasia (the nail–patella syndrome). The *Rh* locus is linked to one variety of hereditary elliptocytosis. The genes for the Lutheran blood group [the only blood group system in which a minus genotype, that is, Lu(a—b—), is inherited as a dominant] and for secretors are closely linked, and are a good example of autosomal linkage.

The first association of blood groups with disease was found in 1953 when it was shown that blood group A was statistically higher in patients with carcinoma of the stomach than in normals. Peptic ulcers were found more commonly in group O individuals than expected by random chance. The incidence of nonsecretors in peptic ulcers was found to be much higher than in a control population; in fact the figures were 50% higher in nonsecretors as compared with secretors. These facts have been confirmed many times by workers all over the world. There are many other associations in the literature that have not been studied so thoroughly.

One of the most statistically significant associations has been the finding of an association between ABO blood groups and secretor status with a polymorphism of serum alkaline phosphatases. Human serum alkaline phosphatases can be shown to be heterogeneous by means of starch gel electrophoresis. Two fractions can be separated: one, rapidly migrating, present in all normal sera, is thought to originate from the liver; the other, with a slower mobility, is thought to originate from the small intestine. The slower fraction was shown to occur almost exclusively in type B and O secretors. It was originally said to be never found in type A secretors or in nonsecretors of any type. Later, more sensitive techniques showed small amounts present in 10 to 15% of A secretors and in a small number of nonsecretors. It was hypothesized that the *ABO* and secretor genes influence (1) the rate at which intestinal alkaline phosphatase enters the bloodstream or (2) the catabolism rather than the synthesis of alkaline phosphatase. It is possible that fat handling is influenced by the *ABO* and secretor genes. Serum cholesterol levels are higher in type A or nonsecretors; thus there is an inverse correlation between ABO-associated factors for cholesterol and intestinal alkaline phosphatase. Studies were initiated to see if these factors are relevant to coronary heart disease, and the initial studies indicate that there may be an association, for in two studies to date there were statistically more group A patients with heart disease.

Other recent studies with relevance to earlier work are the findings in a study on the effects of oral contraceptives in women that the incidence of thromboembolic disease is statistically higher in group A women. Also the coagulation factor VIII is lower in group O as compared with other groups. It has been suggested that this finding may relate to the higher incidence of thrombosis in group A and the bleeding ulcers in group O.

It also appears that the frequency of gastric carcinoma is higher in group A individuals. Many have suggested that one or another blood group might confer selective advantage in combating infection and that the present-day blood group gene frequencies are related to bacterial epidemics that occurred in the distant past before the advent of antibiotics. Indeed recent studies have shown that individuals of blood groups B and AB have a significantly increased susceptibility to serious gram-negative enteric pathogens (other than *Shigella*). This was especially true for *E. coli* and *Salmonella* species in Negro and Puerto Rican ethnic groups.

A recent study of 600 patients suffering from carcinoma showed abnormalities and frequent deviations of antigens of the P blood group system in cases of carcinoma of the bronchi. P_1 was found decreased and P_2 correspondingly increased in these patients, the difference being significant ($\chi^2 = 6.23$). No other correlation between any of the investigated blood group traits and carcinoma could be shown.

The risk of choriocarcinoma's developing after any form of pregnancy appears critically related to the ABO groups of both the woman and her husband. Women of group A married to men of group O seem to have the highest risk, whereas women of group A married to men of group A have the lowest: the relative risk of these two groups was found to be 10.4 to 1. The spontaneous regression of trophoblast after evacuation of hydatidiform mole happened most commonly in women mated to men of their own ABO phenotype. Group AB patients tend to have rapidly progressive tumors that do not respond well to chemotherapy. The data indicate that genetic factors (which are possibly mediated immunologically) strongly affect the development of choriocarcinoma.

Another immunological aspect of neoplasia associated with blood group antigen is the association of breast cancer with lymphoproliferative disease, which is unusual. An association rate of 1.5% was found in 387 radical mastectomies (Hermann, 1971). The six breast cancer patients with associated lymphoma or leukemia all had blood type A, which may indicate a more likely association of these neoplasms in type A patients.

In anthropological studies, blood group markers have proved useful, as in tracing the migrations (and presumably the ethnic origin) of various inbred populations. For example, the frequency of Rh_o-negative blood is approximately 15% in Caucasians, 3 to 7% in African Negroes, 1% in Chinese, and absent in Melanesians. It is about 35% in the Basques of Spain. Among the Rh antigens the E^T antigenic variant is very common in Caucasians but frequently absent as a marker in Australian aborigines.

The Diego (Di^a) blood group has proved especially useful in such studies. In studies of Mongolian and South American populations 1,000 Venezuelans were all Di(a—), whereas 36% of samples from certain South American Indians were Di(a+) and 5 to 15% of Japanese and Chinese had the antigen.

The P_1 alleles have been very useful in similar studies of racial migration. Most Europeans are P_1 (78%), but the frequency is lower in Finland (about 70%). West Africans are more than 95% positive, Japanese 21%.

RECOMMENDED READINGS

Books

Delaney, J. W. and Garratty, G. (1969), *Handbook of Haematological and Blood Transfusion Technique* (2nd ed.). New York: Apple-Century-Crofts.

Giblett, E. R. (1969), *Genetic Markers in Human Blood*. Philadelphia: F. A. Davis Co.

Issitt, Peter D. (1970), *Applied Blood Group Serology*. Oxnard, Calif.: Becton, Dickinson & Co.

Kabat, E. A. (1956), *Blood Group Substances, Their Chemistry and Immunochemistry*. New York: Academic Press.

Klein, J. (1982), *Immunology, The Science of Self–Nonself Discrimination* (1st ed.). New York: Wiley, pp. 156–255.

Mollison, P. L. (1961, 1967), *Blood Transfusion in Clinical Medicine* (4th ed.), Philadelphia: F. A. Davis Co.

Mourant, A. E. (1964), *The Distribution of the Human Blood Groups*. Oxford: Blackwell.

Race, R. R. and Sanger, R. (1968), *Blood Groups in Man* (5th ed.). Philadelphia: F. A. Davis Co.

Race, R. R. and Sanger, R. (1975), *Blood Groups in Man* (6th ed.). Philadelphia: F. A. Davis Co.

Terasaki, P. I. (Ed.) (1982), *Blood Transfusion and Transplantation*. New York: Grune & Stratton.

Zmijewski, C. M. (1978), *Immunohematology* (3rd ed.). New York: Appleton-Century-Crofts.

Reviews

Ginsburg, V., Kabata, A., Hickey, C., and Sawicka, T. (1971), Biochemical basis for blood types in man. In G. A. Jamieson and J. Tibor (Eds.), *Glycoproteins of Blood Cells and Plasma*. Philadelphia: Lippincott.

Laine, R. A., Stellner, K., and Hakomori, S. I. (1974), Isolation and characterization of membrane glycosphingolipids. In E. D. Korn (Ed.), *Methods in Membrane Biology* (Vol. 2). New York: Plenum, p. 205.

Lemieux, R. U. (1978), Human blood groups and carbohydrate chemistry. *Chem. Soc. Rev.* 7, 423.

Morgan, W. T. J. and Watkins, W. M. (1969), Genetic and biochemical aspects of human blood group A-, B-, H-, Le^{a}- and Le^{b}- specificity. *Br. Med. Bull.* 25, 30.

Springer, G. F., Desai, R. F., Yang, H. J., and Murthy, M. S. (1977), Carcinoma-associated blood group MN precursor antigens against which all humans possess antibodies. *Clin. Immunol. Immunopathol.* 7, 426.

Watkins, W. M. (1980), Biochemistry and genetics of the ABO, Lewis, and P blood group systems. *Advances in Human Genetics* 10, 1.

Articles

Adams, J., Broviac, M., Brooks, W., Johnson, N. R., and Issitt, P. D. (1971), *The Probable First Example of Anti-*WI^{b}. Presented at the 24th annual meeting of the American Association of Blood Banks, Chicago, Sept. 12–16.

Albrey, J. A., Vincent, E. E. R., Hutchinson, J., Marsh, W. L., Allen, F. H., Jr., Gavin, J., and Sanger, R. (1971), A new antibody, anti-Fy3, in the Duffy blood group system. *Vox Sang.* 20, 21.

Allen, F. H., Lewis, S. J., and Fudenberg, H. H. (1958), Studies of anti-Kp^{b}, a new antibody in the Kell blood group system. *Vox Sang.* 3, 1.

Allen, F. H. and Rosenfield, R. E. (1961), Notation for the Kell blood group system. *Transfusion* 1, 305.

Aminoff, D., Morgan, W. T. J., and Watkins, W. M. (1950), The isolation and properties of the human blood group A substance. *Biochem. J.* 46, 426.

Andresen, P. H. (1948), Blood groups with characteristic phenotypical aspects. *Acta Pathol. Microbiol. Scand.* (A) 24, 616.

Barrasso, C., Eska, P., Grindon, A. J., Oyen, R., and Marsh, W. L. (1975), Anti-K18: an antibody defining another high-frequency antigen related to the Kell blood group system. *Vox Sang.* 29, 124.

Blumenfeld, O. O. and Adamany, A. M. (1978), Structural polymorphism within the amino-terminal region of MM, NN, and MN glycoproteins (glycophorins) of the human erythrocyte membrane. *Proc. Natl. Acad. Sci.* (*USA*) 75, 2727.

Boettcher, B. (1978), Sequence of action of genes at the secretor, H, ABO and Lewis loci. *Hum. Hered.* 28, 426.

Boston Collaborative Drug Surveillance Program (1971), Relation between cancer and S blood antigen system. *Lancet* 1, 301.

Cartron, J. P., Reyes, F., Gourdin, M. F., Garrettu, M., and Salmon, C. (1977), Antigen site distribution among "weak A" red cell populations. *Immunology* 32, 233.

Cartron, J. P., Ropars, C., Calkouska, Z., and Salmon, Ch. (1976), Detection of A_1A_2 and A_2m^{A1} heterozygotes among human A blood group phenotypes. *J. Immunogenet.* 3, 155.

Ceppellini, R. (1955), On the genetics of secretor and Lewis characters: a family study. *Proceedings of the 5th International Congress of Blood Transfusion*, Paris.

Chester, M. A. and Watkins, W. M. (1969), Alpha-L-fucosyl transferases in human submaxillary gland and stomach tissues associated with the H, Le^a and Le^b blood group characters and ABH secretor status. *Biochem. Biophys. Res. Commun.* 34, 835.

Childs, R. A., Kapadia, A., and Feizi, T. (1980), Expression of blood group I and its active carbohydrate sequences on cultured human and animal cell lines assessed by radioimmunoassays with monoclonal cold agglutinins. *Eur. J. Immunol.* 10, 379.

Chown, B., Lewis, M., and Kaita, H. (1957), A new Kell blood group phenotype. *Nature* 180, 711.

Coombs, R. A., Mourant, R. E., and Race, R. R. (1946), *In vivo* isosensitization of red cells in babies with haemolytic disease. *Lancet* 1, 26.

Cory, H. T., Yates, A. D., Donald, A. S. R., Watkins, W. M., and Morgan, W. T. J. (1974), The nature of the human blood group P_1 determinant. *Biochem. Biophys. Res. Commun.* 61, 1289.

Donahue, R. P., Beas, W. B., Renwick, J. H., and McKusick, V. A. (1968), Probable assignment of the Duffy blood locus to chromosome 1 in man. *Proc. Natl. Acad. Sci.* (*USA*) 61, 949.

Dunsford, I. (1962), cited by Race and Sanger (1962), *Blood Groups in Man* (4th ed.). Oxford: Blackwell, p. 313.

Feizi, T. (1980), The monoclonal antibodies of cold agglutinin syndrome. *Med. Biol.* 58, 300.

Feizi, T., Childs, R. A., Watanabe, K., and Hakonori, S. I. (1979), Three types of blood group I specificity among monoclonal anti-I autoantibodies revealed by analogues of a branched erythrocyte glycolipid. *J. Exp. Med.* 149, 975.

Feizi, T., Kabat, E. A., Vicari, G., Anderson, B., and Marsh, W. L. (1971), Immunochemical studies on blood groups. XLVII. The I antigen complex—precursors in the A, B, H, Le^a and Le^b blood group system. Hemagglutination-inhibition studies. *J. Exp. Med.* 133, 39.

Feizi, T., Kabat, E. A., Vicari, G., Anderson, B., and Marsh, W. L. (1971), Immunochemical studies on blood groups. XLIX. The I antigen complex: specificity differences among anti-I sera revealed by quantitative precipitin studies; partial structure of the I determinant specific for one anti-I serum. *J. Immunol.* 106, 1578.

Fisher, R. A., cited by Race, R. R. (1944), An "incomplete" antibody in human serum. *Nature* 153, 771.

Fisk, R. T. and Foord, A. G. (1942), Observations of the Rh agglutination of human blood. *Am. J. Clin. Pathol.* 12, 545.

Friedenreich, V. (1931), Über die serologie der untergruppen A1 and A2. *Z. Immunoforsch.* 71, 283.

Friedenreich, V. (1936), Eine bisher unbekannte Blutgruppeneigenschaft. *Z. Immunoforsch.* 89, 409.

Fudenberg, H. H., Kunkel, H. G., and Franklin, E. C. (1959), High molecular weight antibodies. *Acta Haematol. (Basel)*, Fasc. 10, 522.

Garratty, G., Hafleigh, B., Dalziel, J., and Petz, L. D. (1971), *A Unique Example of anti-I^T*. Presented at the 24th annual meeting of the American Association of Blood Banks, Chicago, Sept. 12–16.

Gilbert, E. R. (1961), A critique of the theoretical hazard of inter- vs. intraracial transfusion. *Transfusion* 1, 233.

Giblett, E. R., Klebanoff, S. J., Pincus, S. H., Swanson, J., Park, B. H., and McCullough, J. (1971), Kell phenotypes in chronic granulomatous disease: a potential transfusion hazard. *Lancet* 1, 1235.

Goldstein, J., Sivglia, G., Hurst, R., Lenny, L., and Reich, L. (1982), Group B erythrocytes enzymatically converted to group O survive normally in A, B, and O individuals. *Science* 215, 168.

Grollman, A. P. and Marcus, D. M. (1966), Enzymatic incorporation of fucose into blood group H substance. *Biochem. Biophys. Res. Commun.* 25, 542.

Grollman, E. F. and Ginsburg, V. (1967), Correlation between secretor status and the occurrence of 2′-fucosyl-lactose in human milk. *Biochem. Biophys. Res. Commun.* 28, 50.

Grollman, E. F., Kobata, A., and Ginsburg, V. (1969), An enzymatic basis for Lewis blood types in man. *J. Clin. Invest.* 48, 1289.

Grubb, R. (1948), Correlation between Lewis blood group and secretor character in man. *Nature* 162, 933.

Grubb, R. (1951), Observations on the human group system Lewis. *Acta Pathol. Microbiol. Scand.* (*B*) 28, 61.

Grubb, R. and Morgan, W. T. J. (1949), The "Lewis" blood group characters of erythrocytes and body fluids. *Br. J. Exp. Pathol.* 30, 198.

Hakomori, S. I. (1970), Glycosphingolipids having blood group ABH and Lewis specificities. *Chem. Phys. Lipids* 5, 96.

Hasekura, H., Ishimori, T., Furusawa, S., Kawaguchi, H., Kawada, K., Shishido, H., Komiya, M., Fukuoka, Y., and Miwa, S. (1971), Hematological observations on the rh (—/—) propositus, the homozygote of amorphic Rh blood group genes. *Proc. Japan Acad.* 47, 579.

Hearn, V. M., Smith, Z. G., and Watkins, W. M. (1968), An *N*-acetyl-D-galactosaminyl transferase associated with the human blood group A character. *Biochem. J.* 109, 315.

Hermann, J. B. (1971), Lymphoproliferative disease secondary to breast cancer. *N.Y. State J. Med.* 71, 1108.

Hughes-Jones, N. C., Green, E. J., and Hunt, V. A. M. (1975), Loss of Rh antigen activity following the action of phospholipase A2 on red cell stroma. *Vox Sang.* 29, 184.

Kabat, E. A., Liao, J., Burzynska, M. H., Wong, T. C., Thogersen, H., and

Lemieux, R. U. (1981), Immunochemical studies in blood groups. LXIX. The conformation of the trisaccharide determinant in the combining site of anti-Ma (Group 1). *Mol. Immunol.* 18, 873.

Kabat, E. Z. and Leskowitz, S. (1955), Immunochemical studies on blood groups. (XVII) structural units involved in group A and B specificity. *J. Am. Chem. Soc.* 77, 5159.

Kirschbaum, J. and Springer, G. F. (1975), Subunit composition of human blood group MN antigens. *Immunochemistry* 12, 653.

Kisalius, E. C. and Kabat, E. A. (1978), Immunochemical studies on blood groups. LXVI. Competitive binding assays of A1 and A2 blood group substances with insolubilized anti-a serum and insolubilized a agglutinin from *Dolichos biflorus*. *J. Exp. Med.* 147, 830.

Kobata, A. and Ginsburg, V. (1970), Uridine diphosphate-N-acetyl-D-galactosamine: D-galactose α-3-N-acetyl-D-galactosaminyltransferase, a product of the gene that determines blood type A in man. *J. Biol. Chem.* 245, 1484.

Kobata, A., Grollman, E. F., and Ginsburg, V. (1968), An enzymatic basis for blood type A in humans. *Arch. Biochem. Biophys.* 124, 609.

Kogure, T. (1975), The action of group Bm or cisAB sera on group O red cells in the presence of UDP-D-galactose. *Vox Sang.* 29, 51.

Kornstad, L. and Heisto, H. (1957), The frequency of formation of Kell antibodies in recipients of Kell-positive blood. In Aa. Videbaek (Ed.), *Proceedings of the Sixth Congress of the European Society of Haematology*. Basel: Karger, p. 754.

Landsteiner, K. (1901), Über Agglutination-serscheinungen normalen menschlichen Blutes. *Klin. Wochenschr.* 14, 1132.

Landsteiner, K. and Levine, P. (1927), Further observations on individual differences in human blood. *Proc. Soc. Exp. Biol. Med.* 24, 941.

Landsteiner, K. and Weiner, A. S. (1940), An agglutinable factor in human blood recognizable by immune sera for Rhesus blood. *Proc. Soc. Exp. Biol. Med.* 43, 223.

Lanset, S. and Ropartz, C. (1971), A second example of acquired B-like antigen in a healthy person. *Vox. Sang.* 20, 82.

Lau, F. O. and Rosse, W. F. (1975), The reactivity of red blood cell membrane glycophorin with "cold-reacting" antibodies. *Clin. Immunol. Immunopathol.* 4, 1.

Lemieux, R. U., Balser, D. A., Weinstein, W. M., and Switzer, C. M. (1981), Artificial antigens. Antibody preparations for the localization of Lewis determinants in tissues. *Biochemistry* 20, 199.

Lenhard, V., Seelig, A. P., Geisen, H. P., and Roelcke, D. (1978), Identification of *I/i*, *Ph1–3*, and *Gd* antigens in the human kidney: possible relevance to hyperacute graft rejection induced by cold agglutinins. *Clin. Exp. Immunol.* 33, 276.

Levine, P., Backer, M., Wigod, M., and Ponder, R. (1949), A new human hereditary blood property (Cellano) present in 99.8% of all bloods. *Science* 109, 464.

Levine, P., Burnham, L., Latzen, E. M., and Vozel, P. (1941), The role of iso-immunization in the pathogenesis of erythroblastosis fetalis. *Am. J. Obstet. Gynecol.* 42, 925.

Levine, P. and Celano, M. J. (1967), Agglutinizing specificity for LW factor in guinea pig and rabbit anti-Rh serum. *Science* 156, 1744.

Levine, P., Celano, M. J., Vos, G. H., and Morrison, J. (1962), The first human blood –/–, which lacks the "D-like" antigen. *Nature* 194, 304.

Levine, P., Celano, M. J., Wallace, J., and Sanger, R. (1963), A human D-like antibody. *Nature* 198, 596.

Lloyd, K. O. and Kabat, E. A. (1968), Immunochemical studies on blood groups. XLI. Proposed structures for the carbohydrate portions of blood group A, B, H, Lewis^a and Lewis^b substances. *Proc. Natl. Acad. Sci.* (*USA*) 61, 1470.

Lloyd, K. O., Kabat, E. A., Layug, E. J., and Gruezo, F. (1966), Immunochemical studies on blood groups. XXXIV. Structures of some oligosaccharides produced by alkaline degradation of blood group A, B, and H substance. *Biochemistry* 5, 1489.

Marcus, D. M., Kundu, S. K., and Suzuki, A. (1981), The P blood group system: recent progress in immunochemistry and genetics. *Semin. Hematol.* 18, 63.

Marsh, W. L. (1977), The Kell blood group, *Kx* antigen, and chronic granulomatous disease. *Mayo Clin. Proc.* 52, 150. (a)

Marsh, W. L. (1977), Mapping assignment of the Rh and Duffy blood group genes to chromosome 1. *Mayo Clin. Proc.* 52, 145. (b)

Marsh, W. L., Chaganti, R. S. K., German, J., Seidl, S., and Speilmann, W. (1974), Cytogenetic investigation of the Rh_{null} phenotype. *Vox Sang.* 27, 190.

Marsh, W. L., Oyen, R., Nichols, M. E., and Allen, F. H. (1975), Chronic granulomatous disease and the Kell blood groups. *Br. J. Haematol.* 29, 247.

Marsh, W. L., Uretsky, S. C., and Douglas, S. D. (1975), Antigens of the Kell blood group system on neutrophils and monocytes: their relation to chronic granulomatous disease. *J. Pediatr.* 87, 1117.

Mays, W. R. and Pausch, V. (1976), RB, a determinant defined by lymphocyte toxicity being associated with ABO blood groups and ABH secretor status. *J. Immunogenet.* 3, 367.

Moore, B. P. L. (1976), Classification of weak variants of blood group A. *Rev. Fr. Transfus. Immunohematol.* 19, 25.

Morton, J. A., Pickles, M. M., and Sutton, L. (1969), The correlation of the Bg^a blood group with the HLA-7 leucocyte group: demonstration of antigenic sites on red cells and leucocytes. *Vox Sang.* 17, 536.

Mourant, A. E. (1946), A "new" human blood group antigen of frequent occurrence. *Nature* 158, 237.

Naiki, M., Fong, J., Ledeen, R., and Marcus, D. M. (1975), Structure of the human erythrocyte blood group P1 glycosphingolipid. *Biochemistry* 15, 4831.

Naiki, M. and Marcus, D. M. (1975), An immunochemical study of the human blood group P1, P, and p^k glycosphingolipid antigens. *Biochemistry* 14, 4837.

Nakajima, H. and Yokota, T. (1977), Two Japanese families with p^k members. *Vox Sang.* 32, 56.

Oriol, R. (1980), Interactions of ABO, Hh, Secretor and Lewis systems. *Blood Transfus. Immunohematol.* 23, 517.

Oriol, R., Danilovs, J., and Hawkins, B. R. (1981), A new genetic model proposing that the *Se* gene is a structural gene closely linked to the *H* gene. *Am. J. Hum. Genet.* 33, 421.

Pardoe, G. I., Uhlenbruck, G., and Reisenberg, U. (1971), Structural studies of red cells: immunochemistry of the *M,N* and related antigens of the myxoviral receptors. *Med. Lab. Tech.* 28, 255.

Plato, C. C., Mann, J. D., Schwartz, J. T., and Leyshon, W. (1970), The role of Xg^a blood group incompatibility in fetal loss. *Acta Genet. Med. Gemellol. (Rome)* 19, 235.

Pollack, W., Hager, H. J., Reckel, R., Toren, D. A., and Singher, H. O. (1965), A study of the forces involved in the second stage of hemagglutination. *Transfusion* 5, 158.

Riches, R. A. and Laycock, C. M. (1980), A new low frequency antigen Li^a (Livesey). *Vox Sang.* 38, 305.

Ritter, H., Lariverdian, G., Arnold, H., Blume, K. G., Schröter, W., Zimmerschitt, E., Brittinger, G., Konig, E., and Wendt, G. G. (1971), Evidence for linkage between the locus for the ABO system and the locus for phosphoglucose-isomerase (PGI). *Humangenetik* 11, 349.

Robinson, M. G., Tochin, D., and Holpern, C. (1971), Enteric bacterial agents and the ABO blood groups. *Am. J. Hum. Genet.* 23, 135.

Rosenfield, R. E., Allen, F. H., Jr., Swisher, S. N., and Kochwa, S. (1962), A review of Rh serology and presentation of a new terminology. *Transfusion* 2, 287.

Salmon, Ch. (1976), ABO antigens and enzymes. As an introduction. *Rev. Fr. Transfus. Immunohematol.* 19, 2.

Sanger, R. (1955), An association between the P and Jay systems of blood groups. *Nature* 176, 1163.

Seaman, M. J., Benson, R., Jones, M. N., Morton, J. A., and Pickles, M. M. (1967), The reaction of the Bennett–Goodspeed group of antibodies tested with the AutoAnalyzer. *Br. J. Haematol.* 13, 464.

Shen, L., Grollman, E. F., and Ginsburg, V. (1968), An enzymatic basis for secretor status and blood group substance specificity in humans. *Proc. Natl. Acad. Sci. (USA)* 59, 224.

Singal, D. P., Blajchmen, M. A., Naipaul, N., and Heddle, N. (1981), HLA-restricted responses to M and N determinants in the primed lymphocyte test. *Hum. Immunol.* 3, 325.

Skov, F. (1976), Observations of the number of available G (rh^GRh12) antigen sites on red cells. *Vox Sang.* 31, 124.

Springer, G. F. and Ansell, N. J. (1958), Inactivation of human erythrocyte agglutinogens M and N by influenza viruses and receptor-destroying enzyme. *Proc. Natl. Acad. Sci. (USA)* 44, 182.

Springer, G. F., Nagai, Y., and Tegtmeyer, H. (1966), Isolation and properties of human blood-group NN and meconium-Vg antigens. *Biochemistry* 5, 3254.

Stroup, M., MacIlroy, M., Walker, R., and Aydellotte, J. V. (1965), Evidence that Sutter belongs to the Kell blood group system. *Transfusion* 5, 309.

Sussman, L. N., Solomon, R., and Grogan, S. (1975), The Kidd minus-minus phenotype in a Caucasian. *Transfusion* 15, 356.

Toma, V. A., Itarralde, M., Anderson, J. D., and Retief, F. P. (1979), Is the d an alleleic gene of D in the Rh blood group system? A quantitative study using ^{99m}Technetium-pyrophosphate-labelled antibody (TcLA) method. *J. Immunol. Methods* 27, 273.

Voak, D., Cawley, J. C., Emmines, J. P., and Barker, C. R. (1974), The role of enzymes and albumen in haemagglutination reactions. A serological and ultrastructural study with ferritin-labelled anti-D. *Vox Sang.* 27, 156.

Voak, D. and Lodge, T. W. (1971), The demonstration of anti-HI/Hi-H blood group determinants with branched cores. *Nature* 262, 702.

Vos, G. H. (1966), The serology of anti-Tj^a-like hemolysins observed in the serum of threatened aborters in Western Australia. *Acta Haematol.* 35, 272.

Vos, G. H., Celano, M. J., Falkowski, F., and Levine, P. (1961), Relationship of a hemolysin resembling anti-Tj^a to threatened abortion in Western Australia. *Transfusion* 4, 87.

Vos, G. H. and Kirk, R. L. (1962), A "naturally-occurring" anti-E which distinguishes a variant of the E antigen in Australian aborigines. *Vox Sang.* 7, 22.

Vos, G. H. and Moores, P. P. (1976), Hemagglutination inhibition studies of water soluble blood group substances recovered from the erythrocytes of classical Bombay Oh subjects. *Transfusion* 16, 421.

Vos, G. H., Petz, L. D., and Fudenberg, H. H. (1971), Specificity and immunoglobulin characteristics of acquired hemolytic anemia autoantibodies. *J. Immunol.* 106, 1172.

Watkins, W. M., Yates, A. D., Greenwell, P., Bird, G. W. G., Gibson, T. C. F., Wingham, J., and Loeb, W. (1981), A human dispermic chimera first suspected from analyses of the blood group gene-specified glycosyltransferases. *J. Immunogenet.* 8, 113.

Weiner, W. and Vos, G. H. (1963), Serology of acquired hemolytic anemias. *Blood* 22, 606.

Wiener, A. S. and Sonn, E. V. (1946), The Rh series of genes with special reference to nomenclature. *Ann. N.Y. Acad. Sci.* 46, 969.

Yamaguchi, H., Okubo, Y., Seno, T., Matsushita, K., and Daniels, G. L. (1979), A "new" allele, Kp^c, at the Kell complex locus. *Vox Sang.* 36, 29.

Yamashita, K., Tachibana, Y., Takasaki, S., and Kobata, A. (1976), ABO blood group determinants with branched cores. *Nature* 262, 702.

Yoshida, A., Davé, V., Allan, J., and Morel, P. A. (1980), Blood group glycosyltransferase activities in plasma from blood chimera subjects. *Acta Haematol.* 64, 230.

CHAPTER 7

Genetically Determined Immune Deficiency States

The ability to combat foreign infectious agents that invade our tissues is necessary for survival. It has been recognized only recently that even bacteria possess enzymes (restriction endonucleases) that can function to eliminate dangerous foreign DNA of other species. Human ability to counteract microbial invasiveness resides to great degree in a phagocytic, cell-mediated, and humoral immunoresponsiveness. The components of the immune response must act in balanced concert; inherited disorders in any of these components predispose affected individuals to great risk. Since the different limbs of the immune response must act together, it is to be expected that the resistance they confer be on a multigenic basis. Understanding of the genetic factors involved in the immune response to a particular pathogen has been expedited by the development of genetically inbred animal strains. Thus one inbred murine line might be highly susceptible to ectromelia virus, yet equally resistant to *Salmonella;* in the second inbred line, the relationship may be reversed. In addition, certain "experiments of nature" (naturally occurring immunodeficiency states in humans) and natural animal models of immunodeficiency have thrown much light on the genetic mechanisms that underlie the normal functioning of the components of the immune response.

CLINICAL SYNDROMES ASSOCIATED WITH PHAGOCYTE DYSFUNCTION

Cells of the monocyte–macrophage series are derived from a different hematopoietic stem cell lineage than the lymphocytic stem cell. Table 7.1 depicts some of the most renowned primary phagocytic dis-

Table 7.1 Primary phagocyte defects. (Reprinted from WHO Report on Immunodeficiencies from *Clin. Immunol. Immunopathol.* 13: 296, 1979)

Designation	Functional defect	Affected cells*	Defective mechanism	Inheritance†	Associated feature
Chronic granulomatous disease	Bacterial killing	N, M	Peroxide and superoxide production	X-linked	Mothers have cutaneous lupus erythematosus
Chronic granulomatous disease	Bacterial killing	N, M	Peroxide and superoxide production	AR	
Myeloperoxidase deficiency	Bacterial killing	N	Peroxide and superoxide production	AR	
Leukocyte G6PD deficiency	Bacterial killing	N	Peroxide and superoxide production	?	
Leukocyte pyruvate kinase deficiency	Bacterial killing	N	?	?	
Chediak–Higashi disease	Mobility and bacterial killing	N, M	?	?	Hair pigment change, granules in phagocytes
Actin-binding deficiency	Mobility	N	Actin binding	AR	
Shwachman's disease	Mobility	N	?	AR	Pancreas malfunction, bone abnormalities, etc.
Others	Mobility	N	?	?	
Defective clearance	Particle clearance	N	?	Polygenic	

* M = macrophage; N = neutrophil.
† AR = autosomal recessive.

orders and the suspect genetic defects. An important mechanism for eliminating particulate foreign matter involves internalization by specialized cells. This process is termed phagocytosis (literally "eating"), and the relevant cells are termed phagocytic cells. In the case of most bacteria the organism is killed and subsequently digested. Consequently, defective function at one or more of the steps comprising the phagocytic process leads inexorably to failure in removing certain microorganisms and thus greatly increased susceptibility to infection. Before discussing clinical syndromes associated with phagocyte dysfunction, it may be helpful to consider normal phagocytic mechanisms. (Humoral and cell-mediated antimicrobial defenses are considered elsewhere in the text.) This discussion will concentrate on polymorphonuclear phagocytes (PMN), but although less is known about mononuclear phagocytes, the mechanisms involved are generally similar. Mention will also be made of some of the more important assays used to test phagocyte function.

Phagocytic Cells

Most cells in the body are theoretically capable of phagocytosis. However, this function devolves predominantly on a group of specialized cells that are "professional" phagocytes. These include PMN and monocytes in the peripheral circulation, and reticuloendothelial macrophages. The latter are thought to derive from monocytes but are fixed in tissues, for example, lining the endothelium of blood vessels in lymph nodes and organs such as the liver (Kuppfer cells) and spleen (littoral cells); they also are present in skin (Langerhan cells), and probably in kidney (mesangial cells), brain (some glial cells), and other tissues as well. Clearly, adequate numbers of these cells are necessary for efficient phagocytosis, and it has been estimated that up to 10 kilograms (20 pounds) of PMN are formed in the bone marrow every day.

Overall View of Phagocytosis

The term phagocytosis is often used to mean internalization of particles. However, several other processes are intimately involved, and these may be considered for convenience under several headings. Thus the first prerequisite is the approximation of phagocytic cells to the invading microorganism, particularly when the latter is situated outside the confines of the peripheral circulation. This requires movement of the phagocytic cell, which in contrast to the random, disorganized movement that occurs spontaneously in such cells, is organized (locomotion) and is directed by chemical attractant signals emanating from the microorganism (chemotaxis). The second major heading is that of opsonization ("buttering" the microorganisms), in which interactions between the cells, antibodies, and complement not only provide chemotactic signals that attract phagocytic cells, but also facilitate attachment of the microorganism to the phagocyte. The latter is a necessary step before the third phase of phagocytosis, which involves engulfment and ingestion of the microorganism. This process stimulates a wide variety of metabolic and morphological changes in the phagocytic cell resulting in the final phase, intracellular killing, destruction, and disposal of the microorganism.

Chemotaxis. The movement of cells outside the vascular compartment requires first that they become attached to the endothelium of blood vessels (a process known as margination) before emigration from the vascular compartment (diapedesis). Margination is tested by measurement of the PMN "stickiness" (in terms of their ability to adhere to nylon wool), and diapedesis is assessed by monitoring the emigration of cells from intact capillaries in the skin after making a bloodless abrasion (Rebuck window). The latter can be done quantitatively by measuring the number of cells appearing as a function of time, or qualitatively, in which case PMN will normally be the predominant cell type seen during the first 6 hours or so, whereas after this the cells most commonly seen will be monocytes.

Many factors mediate chemotaxis. The most important are the complement components and their fragments C3a, C5a, and an activated complex of C5, C6, and C7 ($C\overline{567}$). Other important factors include proteolytic enzymes (from tissue or of bacterial origin), thrombin, which can directly cleave complement components producing chemotactic fragments, and certain components of other sequentially activated systems such as kinins and fibrinolytic mechanisms. Finally, interactions of other cell types, such as lymphocytes with antigen, result in the production of chemotactic factors, as well as other substances—soluble lymphocyte-derived factors with biological activities ("lymphokines"), which activate monocytes (MAF) and cause inhibition of migration of PMN (LIF) and macrophages (MIF). Chemotaxis can be examined *in vitro*. Briefly, the PMN under scrutiny are introduced into the upper compartment of a two-chambered vessel (Boyden chamber) and stimulated to squeeze actively through a membrane separating the upper chamber from the lower chamber, the latter containing serum that has been altered to produce chemotactic signals. The number of cells migrating through the membrane represents the chemotactic index. Chemotaxis may be deficient as a result of either abnormal PMN or reduced ability of serum from the patient under test to generate chemotactic activity.

Opsonization. The precise events involved are ill-understood, but one obvious mechanism is the binding of specific IgG and/or IgM antibody to the microorganism, followed by binding and activation of complement via the classical pathway. This not only generates chemotactic signals (C3a, C5a, and $C\overline{567}$), but also results in binding of the microorganism to phagocytic cells via Fc and C3 receptors on the membrane of the phagocytes. This in effect immobilizes the microbe so that ingestion can occur. Currently, it is thought that C3 receptors are predominantly involved in binding of microorganisms, whereas Fc receptors are necessary for ingestion, but this hypothesis has not yet been conclusively proved. Opsonization also can occur as a result of the alternative pathway of complement activation. Ingestion by phagocytes is further facilitated by a basic tetrapeptide on the phagocyte surface membrane (threonyl-lysyl-prolyl-arginine) termed Tuftsin. As an alternative to binding of the phagocytic cell to the microorganism by immunological means in order to facilitate ingestion, the microorganism may be held still by physical means, for example in the interstices of fibrin clots, a phenomenon known as "surface phagocytosis."

In vitro examination of this phase of phagocytosis involves either testing the opsonizing properties of blood on various microorganisms or assays of serum antibody titers of microorganisms with or without deliberate immunization, and studies of complement.

Ingestion. When firm contact has been established between microorganism and the phagocytic cell, pseudopodia are extruded from the phagocytic cell around the particles. These fuse on the distant side, thus incorporating the particle into the cell within a phagosome, the lining of which is inverted cell membrane. This process is quite different from pinocytosis, a process whereby much smaller substances that are generally undetectable at the light microscopic level are taken up in minute invaginations of the cell membrane. The phagosome buds off from the membrane and moves centripetally.

During this time, fusion is occurring with lysosomal granules to form a phagolysosome. This is associated with the disappearance of lysosomal granules from the cytoplasm (a process known as degranulation). Two populations of granules are recognized: primary and secondary. [Primary granules contain, among other things, myeloperoxidase (MPO), lysosomal enzymes including β-glucuronidase, cationic proteins, and lysozyme. Secondary granules contain lysozyme and alkaline phosphatase.] These processes require energy that, for mobility of the phagocytic cell and ingestion, is largely supplied by glycolytic metabolism.

Ingestion is assessed by measurement of the cellular uptake of particles, with measurement of either the number taken up or the number remaining in the suspending media, both as a function of time. It is important to provide opsonins (antibody and complement), and it should be emphasized that at present no standardized procedure is in general use. Alternatively, inert particles such as latex or oil droplets may be used. One of the most popular methods involves red oil droplets coated with lipopolysaccharide (LPS) opsonized with IgG and complement. After incubation, washing, and cell rupture, the red color is extracted organically and measured spectrophotometrically. As with assays of chemotaxis, the concurrent examination of PMN and serum derived from normal individuals can allow the detection of defects either intrinsic to the cells or in the serum being tested.

Intracellular killing. The multiple mechanisms involved are complex and still somewhat obscure. Broadly, they may be considered in categories of nonoxidative and oxidative. The former mechanisms include the wide variety of substances present in lysosomal granules. For example, the reduced pH of the granules, and the presence of elastase, lipase, and cationic proteins in the phagolysosome may damage the cell membranes of microorganisms. Also, lactoferrin has antimicrobial properties, perhaps related to the chelation of iron essential for microbial metabolism. However, although such mechanisms may play a significant role in the killing of certain microorganisms under anaerobic conditions, they are probably of greater importance in cellular digestion than in killing of bacteria. This is consistent with the fact that mature macrophages lack several of these components but remain capable of intracellular killing. Granules may be examined by light or electron microscopy, and, since fusion with the phagosome may occur before closing and budding off from

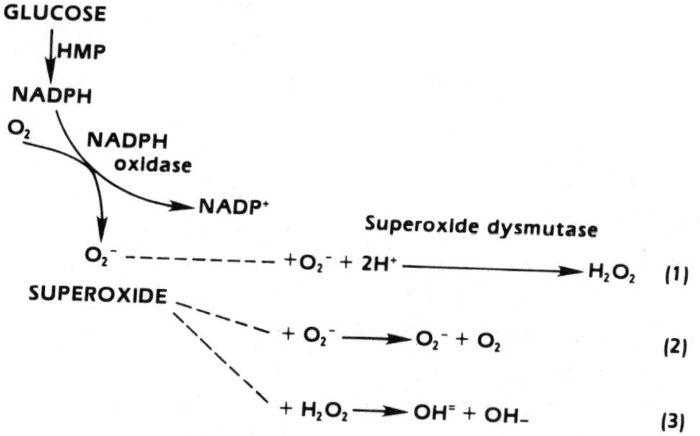

Fig. 7.1 Pathways involved in oxidative intracellular killing.

the membrane, enzymes specific for primary granules (β-glucuronidase and MPO) and for secondary granules (alkaline phosphatase) may be measured in the suspending medium.

Oxidative processes involve several mechanisms, all of which require consumption of oxygen and glucose. One of the most important killing mechanisms requires the generation of hydrogen peroxide (H_2O_2). This is predominantly achieved through the action of NADPH oxidase, an enzyme that lines the phagosome. The pathways involved are complex and will not be considered here in detail; they are summarized in semidiagrammatic form in Figure 7.1.

In contrast to chemotaxis and ingestion, the energy requirements for intracellular killing are supplied almost entirely by metabolism through the hexose monophosphate (HMP) shunt. Glucose metabolism via the HMP shunt generates large quantities of NADPH, which donates an electron to an oxygen molecule in the presence of NADPH oxidase to form superoxide (O_2^-). Although this compound is extremely toxic, it probably does not in itself exert an important bactericidal effect. It may be converted into singlet oxygen and hydroxyl radicals (see reactions 2 and 3 in Fig. 7.1), but the most significant reaction is with the H^+ ions that abound in the acid milieu of the phagolysosome. In the presence of superoxide dysmutase, H_2O_2 is formed (reaction 1). H_2O_2 is involved in a number of microbial processes, as summarized below.

H_2O_2 + MPO + Cl— (halide)
H_2O_2 + lysosyme + acrobic acid + Co^{2+} or Cu^{2+}
H_2O_2 + membrane fatty acids ⟶ malonyldialdehyde

H_2O_2 is itself extremely toxic to host tissues, and at least two enzyme systems apparently function to destroy any H_2O_2 that leaks out of the phagolysosome. The first involves glutathione reductase and the second catalase.

Perhaps not surprisingly, many assays are available to assess this phase of phagocytosis. The ultimate test is the efficiency of intracellular killing of a particular microorganism. Assays of intracellular killing are similar to those that test ingestion, with the exception that the number of viable intracellular bacteria rather than the total number ingested is measured. This is accomplished by causing lysis of PMN, plating the lysate on agar, and counting the number of colonies forming after different time periods. It is important to be certain that ingestion is normal, since apparently normal reduction in the number of colonies on incubation may be due to poor internalization. The postingestion respiratory burst stimulated by ingestion of particles *in vitro* is often assessed by the NBT (nitroblue tetrazolium) test.* This test may be positive in the absence of stimulation by ingestion of particles *in vitro*, as seen in many infections, therapy with certain drugs, and some constitutional disorders. Other assays examine individual components of the respiratory burst, for example the uptake of oxygen (using an oxygen electrode) and of glucose by quantitative evolution of $^{14}CO_2$ from [^{14}C]glucose. The HMP shunt pathway may also be further investigated by measuring the activities of the various enzymes involved.

Clinical Disorders—General Considerations

In a process as complex as phagocytosis there is considerable room for error; thus it is not surprising that defective function of any one of the components may lead to increased susceptibility to severe, recurrent life-threatening infection. For example, the presence of decreased numbers of normally functioning PMN or monocytes, as seen in various hematological disorders, including leukemia, will result in suboptimal phagocytosis. Margination or diapedesis may also be impaired, and several disorders are associated with defective chemotactic ability of PMN, one example being the "lazy leukocyte syndrome." Similarly, immunodeficiency resulting from defective antibody production or from abnormalities in the complement system is

* NBT is a yellow substance that is reduced to a deep blue compound, formazan. This process can be quantitated spectrophotometrically or assessed under the microscope in terms of the percentage of cells that are positive.

also associated with a marked increase in the frequency and/or severity of infection.

Many of these aspects of phagocytic function are affected by a wide variety of constitutional disorders. For example, chemotaxis may be markedly impaired by alcohol and by other "toxins." However, the prime example of impairment of chemotaxis is in patients with diabetes mellitus, who have been reported to have measurable defects at every phase of phagocytosis—from margination, diapedesis, and chemotaxis, through opsonization, to ingestion and intracellular killing. Since many of the necessary steps require energy derived from the metabolism of glucose, it is perhaps not surprising that the disruption of intermediary carbohydrate metabolism that occurs in diabetes mellitus leads to defects in phagocytosis, and that these are, in general, most marked in those patients with the most severe disease (i.e., with hyperglycemia and acidosis). Moreover, these defects provide a logical explanation for the apparently increased frequency of certain types of infections in patients with diabetes mellitus. In addition to these secondary deficiencies of phagocytic dysfunction, several primary disorders have been discovered. These are all rare but are of importance in further understanding normal phagocytic function.

Chronic granulomatous disease (CGD). This is usually inherited as an X-linked recessive disorder and is manifested during the first 2 years of life with recurrent abscesses, lymphadenitis, hepatosplenomegaly, osteomyelitis, and pneumonia. It is now known that all steps of the phagocytic process up to and including ingestion are normal. However, since one or more enzymes involved in the post-ingestion respiratory burst (particularly NADPH oxidase) are deficient, intracellular microbicidal mechanisms are not activated and, consequently, many microorganisms are not killed. This is reflected experimentally in decreased oxygen and glucose uptake, greatly impaired H_2O_2 production, and absence of NBT reduction. In heterozygous carriers of this disorder, NBT is partially reduced, and results are intermediate between those of normal and homozygous individuals. Similarly, intracellular killing is grossly impaired in the case of organisms commonly found in lesions. It is important to note that many of these microorganisms, such as *Staphylococcus epidermidis, Serratia marcescens,* and *Aspergillus* are nonpathogenic in normal individuals. Since the predominant deficiency in CGD is in the production of H_2O_2, relatively normal results for intracellular killing may be obtained in the case of bacteria that can generate H_2O_2 and lack the catalase necessary for its effective catabolism (i.e., streptococci, pneumococci, meningococci, and *Haemophilus influenzae*). This observation also explains the characteristic frequency of non-pathogenic organisms in CGD noted above, since the latter are catalase-positive. Cytochrome b-245 is an heme-containing protein involved in the microbial oxidase system that recently has been shown to be absent from the neutrophils from patients with X-linked CGD (Segal *et al.*, 1983); thus there appear to be two patterns of inheritance of CGD—X-linked and autosomal recessive.

Other enzyme deficiencies. Deficiency of glucose 6-phosphate dehydrogenase (G6PD) causes a similar marked impairment in the generation of H_2O_2 as a consequence of decreased activity of the HMP shunt. The clinical features and susceptibility to microorganisms are similar to CGD, and results of laboratory tests are similar. The diagnosis is most easily made by the finding of decreased G6PD activity in leukocytes. However, this genetic defect is not X-linked but autosomal recessive, and hence it occurs in females as often as in such males.

Chediak–Higashi syndrome. In addition to the increased susceptibility of infection typical of all disorders of phagocytic function, two features are typical of this syndrome—namely, an increased risk of lymphoreticular malignancies and albinism. This disorder can occur in other species. The defect appears to be in the lysosomes, which fuse together and are of giant proportions and functionally ineffective. Consequently oxidative metabolism is normal, but intracellular killing is impaired. In addition, natural killer (NK) cells are decreased and the incidence of malignancy greatly increased. This disease occurs not only in humans but also in Aleutian mink, in certain canine strains, and in mice where it is often associated with albinism, partial or total. Indeed, the senior author maintains that the first report in the literature of this syndrome was that of Herman Melville (*Moby Dick*, 1851).

Job's syndrome. This disorder, an autosomal recessive, was described in 1966 and is so named because of the multiple boils ("cold" abscesses) characteristically present, together with otitis media and sinusitis. Although the abnormality involved is uncertain, the PMN seem unduly susceptible to deactivation, and monocyte chemotaxis is very poor. Job's syndrome is now classified under hyperimmunoglobulin E-recurrent infection (HIE). Esterase-negative mononuclear cells from these patients produce a chemotactic inhibitory factor (molecular weight 61,000) that may partially account for the disorder.

DEVELOPMENTAL DEFECTS OF B CELLS

Our understanding of the cellular basis of the immune response derives in part from the fact that the progenitors of T and B cells arise from a pluripotential common lymphoid cell. The progenitor B-cell gives rise to a rapidly mitotic population of pre-B cells that produce 8S monomeric IgM but lack functional Ig membrane receptors. In humans, this maturation is initiated in the fetal liver and continues in the bone marrow. It is here that the majority of B cells are committed as to which particular Ig genes they will eventually express. The next stage of differentiation culminates in the appearance of IgM on the surface of the immature B lymphocyte, a cell already genetically programmed for the antibody specificity that will later be secreted when it matures into the B cell and terminal plasma cell. These immature B cells are tolerized very easily and thus are pivotal in the expression of Ig-class (isotype) diversity. Next, either IgG or IgA is expressed on a particular subpopulation of $sIgM^+$ cells, and then each of these sublines of cells within a B-cell clone develops surface membrane IgD. Concomitant with this isotype diversification, B cells acquire other cell membrane entities, such as receptors for C3 and the Fc portion of IgG, that enhance interaction with T cells and antigen. It is apparent that at several steps along the maturation pathway, a genetic defect may be reflected in partial or total paralysis of certain immune functions. Interestingly, neither antigens nor T cells are required for early B-cell differentiation; yet the isotype switch from IgM to IgG or IgA (while maintaining the same combining-site specificity) requires antigen and T-cell cooperation.

A deficiency of one or more, but not all, Ig classes is known as a selective Ig deficiency. Those patients with immune deficiencies that resist classification are categorized as having "varied" immunodeficiencies. These include some "congenital" and "acquired" hypogammaglobulinemias. Specific steps in the B-cell differentiation pathway are affected by these defects, and some patients fail to generate sufficient numbers of pre-B cells. Others may show insufficiency in a particular isotype. Thus IgM production may be normal, while there is an accompanying dramatic diminution or absence of B cells with surface IgG or IgA. "Isolated IgA deficiency" exists in two forms. Approximately 1 in 500 normal Caucasians with clinical symptoms have IgA levels less than 2% of normal.

However, a small group of patients exists with such IgA deficiency, lack of secretory IgA, and recurrent gastrointestinal and respiratory infections. In contrast to the normal groups, these individuals lack the IgG2 subclass, and all three genetic forms exist—X-linked recessive, X-linked dominant, and autosomal recessive. Pedigree studies have identified families with deletions of the Ig structural genes for H chains. There are also defects in which circulating B cells are normal, but there is an abnormality associated with B-cell terminal differentiation. For instance, one X-linked trait, the so-called hyper-IgM immunodeficiency syndrome is characterized by elevated serum IgM levels, but a lack of IgG or IgA, even though B cells with corresponding γ- or α-chains are found in the blood. For unknown reasons, the plasma cells are not formed from these latter mature B cells. In some immunodeficiencies characterized by B cells that fail to respond to polyclonal B-cell activators, the defect is intrinsic to the B cell. Some abnormalities in glycosylation exist that affect the secretion of already preformed Ig. Table 7.2 is a list of primary immune deficiency states suggested by the WHO Report on Immunodeficiencies.

DEVELOPMENTAL DEFECTS OF T CELLS

The most clinically prominent element of T-cell deficiencies is increased infection with such agents as fungi, parasites, mycobacteria, and viruses. There is a wide range in the severity of these diseases. For instance, there may be fulminating infections with usually benign agents in severe cell-mediated immunodeficiencies (CMI) to very circumscribed fungal infections in certain antigen-selective T-cell defects. Some degree of graft-versus-host reaction after allogeneic cell transfusion may be the first sign of defective cell-mediated immunity. Also, in patients such as those with familial chronic mucocutaneous candidiasis, lymphoproliferative malignancies are seen much more frequently. Another manifestation of cell-mediated immunity may be autoimmunity with the formation of lymphocytotoxic antibodies that contribute directly to T-cell failure. Thus the spectrum of defective T-cell immunity is expansive and heterogeneous.

On the basis of a more complete comprehension of T-cell matura-

Table 7.2 Classification of primary specific immunodeficiencies. (Reprinted from WHO Report on Immunodeficiencies from *Clin. Immunol. Immunopathol.* 13: 296, 1979)

Designation	Usual phenotypic expression: Functional deficiencies	Usual phenotypic expression: Cellular abnormalities	Presumed level of basic cellular defect*	Known or presumed pathogenetic mechanism	Inheritance†	Main associated features
1. Severe combined ID						
(a) Reticular dysgenesis	CMI, Ab and phagocytes	↓ T, B, and phagocytes	HSC	Unknown	AR	–
(b) "Swiss type"	CMI and Ab	↓ T and B	LSC	Unknown	AR	–
(c) ADA deficiency	CMI and Ab	↓ T ± B	LSC or early T	Metabolic effects of ADA deficiency	AR	± Chondrocyte abnormalities
(d) With B lymphocytes	CMI and Ab	↓ T (B lymphocytes without or with normal isotype diversity)	Early T ± early B	Unknown	X-Linked or AR	
(e) Others						
2. Thymic hypoplasia (DiGeorge's syndrome)	CMI and impaired Ab	↓ T	Thymus	Embryopathy of third and fourth pharyngeal pouch	Usually not familial	(1) Hypopara thyroidism (2) Abnormal facies (3) Cardiovascular abnormalities
3. PNP deficiency	CMI ± Ab	↓ T	T	Metabolic effects of PNP deficiency	AR	Hypoplastic anemia

4. ID with ataxia telangiectasia	CMI and Ab (partial)	$\downarrow$ T and plasma cells (mainly IgA, IgE $\pm$ IgG)	Early T and defective terminal differentiation of B lymphocytes	Unknown ? Faulty thymic epithelium ? DNA repair defect	AR	(1) Cerebellar ataxia (2) Telangiectasia (3) Ovarian dysgenesis (4) Chromosomal abnormalities
5. ID with thymoma	Ab and impaired CMI (variable)	$\downarrow$ Pre-B and B $\pm \downarrow$ T	HSC	Unknown	None	(1) Thymoma (2) Eosinopenia (3) Erythroblastopenia (4) Aplastic anemia
6. X-Linked agammaglobulinemia	Ab	$\downarrow$ B	Pre-B	Unknown	X-Linked	—
7. Transcobalamin II deficiency	Ab and phagocytosis	$\downarrow$ Plasma cells	Failure of terminal differentiation of B lymphocytes	Metabolic effects of vitamin B_{12} deficiency	AR	(1) Pancytopenia with megaloblastic anemia (2) Intestinal villous atrophy
8. Selective IgA deficiency	IgA Ab	$\downarrow$ IgA plasma cells $\pm \uparrow$ Bα lymphocytes $\pm \downarrow$ T	Terminal differentiation of Bα lymphocytes impaired	(1) ? $\uparrow T_S$ (2) ? $\downarrow T_H$ (3) ? Intrinsic B-cell defect	Unknown > AR > AD; frequent in families of patients with varied ID	(1) Occasional 18-chromosomal deletions (2) Anti-IgA antibodies

Table 7.2 (*Continued*)

Designation	Usual phenotypic expression: Functional deficiencies	Usual phenotypic expression: Cellular abnormalities	Presumed level of basic cellular defect*	Known or presumed pathogenetic mechanism	Inheritance†	Main associated features
9. Selective deficiency of one other Ig class or subclass	Ab	↓ Plasma cells ± ↓ T	Unknown	Unknown ? ↑ T_s	Unknown	—
10. Secretory piece deficiency	Secretory IgA Ab	↓ Intestinal IgA plasma cells	Mucosal epithelial cell	Unknown	Unknown	—
11. Ig deficiencies with increased IgM	Ab	↓ IgG and IgA plasma cells, ↑ IgM plasma cells, ± ↑ Bα lymphocytes	Failure of terminal differentiation of Bγ and Bα lymphocytes	Unknown	X-Linked or AR or unknown	—
12. Ig deficiencies with IgM production and without γ and α cells	Ab	Absent Bγ and Bα lymphocytes	Pre-B or B	Faulty isotype diversification	AR or unknown	—
13. Transient hypogamaglobulinemia of infancy	Ab	↓ Plasma cells	Impaired terminal differentiation of B lymphocytes	? ↓ T help	Frequent in heterozygous individuals in families with various severe combined ID	—

14. Antibody deficiency with normal or hypergammaglobulinemia	Impaired Ab for some antigens (mainly primary response)	↓ B	Pre-B or early B	? Reduced clonal size or diversity	AR in some	—
15. κ-Chain deficiency	Ab	↓ Bκ	Pre-B	Unknown	Unknown or familial	—
16. Wiskott–Aldrich syndrome	Ab to certain antigens (mainly polysaccharides) and CMI (progressive)	↓ T and B (progressive)	Unknown	Unknown	X-Linked	(1) Thrombocytopenia (2) Eczema
17. Varied ID (common and largely unclassified)						
(a) Predominant Ig deficiency	Ab ± CMI	± ↓ B	Pre-B or B in some	(1) Intrinsic B-cell defect (2) Underproduction of B cells (3) ? ↑ T_S (4) ? ↓ T_H (5) Autoantibodies to B cells	Unknown or familial	—
(b) Predominant T-cell deficiency	CMI ± Ab	↓ T	Early T or T_H	(1) Unknown (2) Autoantibodies to T cells	Unknown or familial	—

↑↓ = Increase or decrease in level.
* HSC = hematopoietic stem cell; LSC = lymphopoietic stem cell.
† AR = autosomal recessive; AD = autosomal dominant.

tion obtained utilizing differentiation surface markers, clinical immune T-cell deficiencies may be classified into three major categories: (1) failure of prethymic T-cell development, (2) failure of intrathymic maturation, and (3) failure of postthymic T cells (defects of regulation).

Failure of Prethymic T-cell Differentiation

These diseases seem to include all SCID (severe combined immunodeficiency) syndromes. These may include hematopoietic stem cell agenesis or selective T-cell agenesis. Table 7.3 presents a general classification of these disorders, in which they are grouped according to mode of inheritance, associated enzymopathy, and characteristics of circulating lymphocytes.

Reticular Dysgenesis (genetics unclear). This most severe form of SCID is characterized by defects in the development of stem cells and agenesis of both T and B lymphocytes and granulocytes in the bone marrow. Erythro- and thrombopoiesis are usually normal. The failure of cell- and humoral-mediated immunities in concert with phagocytic dysfunction leads to an early death, usually within the first few days of life.

Developmental T- and B-Cell Defects (usually autosomal recessive). This "Swiss type" of SCID is characterized by the absence of lymphoid stem cells. Hematopoietic stem cell development is usually unaffected, and circulating T- and B-cell levels are markedly reduced. The thymus is dysplastic, and the peripheral lymphoid organs are uncolonized. Agammaglobulinemia may be present. Both X-linked and autosomal modes of inheritance have been reported. About one fifth of the patients have a deficiency of the enzyme adenosine deaminase (ADA), inherited as an autosomal recessive. T-cell function is more severely affected than B-cell function in the enzyme deficiency syndromes that may accompany SCID. Some evidence suggests that failure of lymphoid cell development results from toxicity due to high accumulation of adenosine or adenosine nucleotides in those with adenosine deaminase deficiency. This still remains unproved, but the disorder is an example of important experiments of nature. Recently, a few cases of SCID associated with purine nucleophosphorylase (PNP) deficiency have been described. A similar toxic effect of purine metabolites is suggested.

Developmental T-Cell Defects. The "Nezelof syndrome" was first described in a 14-month-old child with recurrent infections resembling SCID, except that serum Ig levels were normal. These patients demonstrate considerable heterogeneity and many show aberrations of humoral immunity. IgM may be present, but IgG and IgA are usually low. Experimental evidence has accrued indicating that the B-cell system may be normal, but lack of T-cell help may contribute to B-cell hypofunction. The basic defect remains unknown. Developmental failure of T-cell progenitors and of thymic factors to induce T-cell maturation have been causally implicated. Bone marrow grafts and thymic transplants have restored normal T- and B-cell function, supporting the validity of the theory.

Failure of Intrathymic Maturation or Thymic Development

A disorder affecting thymic function will ultimately be observed as some impairment in T-cell function. It is obvious that the clinical phenomenology seen in these disorders is complex. Deficiencies in this category may manifest themselves as complete failure of T-cell function (much like Nezelof syndrome), or they may be seen as subtle defects in immunoregulation.

The DiGeorge syndrome (nongenetic, presumably due to intrauterine defects). This syndrome results from abnormal embryogenesis of the third and fourth pharyngeal pouches duirng the first 6–8 weeks of gestation leading to abnormalities of the ears, face, congenital heart disease, and hypoplasia of the parathyroids and thymus. Both parathyroid and thymus loss range from hypoplasia to agenesis. Although total peripheral blood lymphocytes are usually normal in number, the numbers of T cells are markedly reduced. The thymus-dependent areas in peripheral lymphoid organs are depleted, but the extent of *in vitro* and *in vivo* defects in T-cell immunity is heterogeneous. Humoral immunity appears normal, although some patients respond to some T-dependent antigens poorly. Results of therapeutic trials with cultured thymic epithelium suggest that the major pathogenetic defect is the absence of intrathymic T-cell maturation. The children are at high risk of sudden death due to recurrent infection of the upper respiratory tract, diarrhea, and hypocalcemia (unless given calcium infusions and parathyroid and thymic hormones).

Ataxia Telangiectasia (autosomal recessive). Ataxia telangiectasia is characterized by low serum IgA, recurrent upper respiratory tract

Table 7.3 Classification of disorders of lymphoid stem cell differentiation. (Reprint from R. J. O'Reilly *et al.*, *Transplant. Proc.* 10: 187, 1978)

Parameters	SCID with Enzymopathy ADA Deficiency	ADA Inhibitor	SCID with Skeletal Anomalies	SCID (Classical)	SCID With B Cells −Ig	+Ig1	+Ig2
Features							
Inheritance*	A	A	A,S	A,S	A,S	A,S	A
Thymus morphology	Dysplastic	ND†	Embryonal	Embryonal	Embryonal	Embryonal	Hyperplastic, no Hassall's corpuscles
Lymphocyte markers							
B cells (surface I_s)	↓ = N‡	↓ = N	↓	↓	N = ↑	N = ↑	N
T cells (E-rosettes)	↓	↓	↓	↓	↓	↓	↓
B-Cell function							
Antibody production	—	—	—	—	—	—	—
PWM-stimulated plasma cell differentiation	−/+	−/+	ND	—	—	+	ND
PWM-stimulated Ig secretion	−/+	−/+	ND	—	—	+	ND
Supressors of B-cell function					+/−	+/−	ND
T-Cell function							
Lymphocyte transformation							
PHA	↓↓	↓↓	↓↓	↓↓	↓↓	↓↓	↓↓
Allogenic cells	↓↓	↓↓	↓↓	↓↓	↓↓	↓↓	↓↓
Antigens	↓↓	↓↓	↓↓	↓↓	↓↓	↓↓	↓↓
Lymphokine production	—		ND		—	—	ND

(concanavalin A)							
T-Cell helper function	—	ND	ND	—	—	—	—
Delayed hypersensitivity reactions	—		—	—	—	—	—
K-Cell function (ADCC)	↓	↓	ND	↓	↓	N	ND
T-cell differentiation in presence of							
Thymic extract or supernatant							
HTLA+	—/+	—/+	ND	—	+	+	ND
E-Rosettes	—/+	—/+	ND	—	—	—	—
Mitogen responses	—/+	—	ND	—	—	—	—
Thymic epithelium							
HTLA+	—/+	—/+	ND	ND	+	+	ND
E-rosettes	—/+	—/+	ND	ND	+/—	+/—	+
Mitogen responses	—/+	—	ND	ND	+/—	+/—	+
B-Cell differentiation (Ig synthesis and secretion) in presence of normal T lymphocytes	ND	ND	ND	ND	+	—	ND
Thymus–secretory function–thymic hormone							
(Bach, Dardenne)	ND	N	ND	N	↓	N	ND
Thymopoietin (Twomey)	↓	↓	ND	↓	↓	↓	ND

* S = sex-linked recessive; A = autosomal recessive.
† ND = no data.
‡ N = none.

infections, oculocutaneous telangiectasia, often eventual reticuloendothelial disease, and neoplasia. The disease involves the endocrine system as well, including ovarian dysgenesis, testicular atrophy, other endocrine disorders such as pituitary failure, and abnormalities of carbohydrate metabolism. Autoantibodies to insulin receptors may contribute to endocrine disorder. Chronic sinopulmonary infection may culminate in respiratory insufficiency in these patients, and death. Lymphoproliferative malignancies and brain tumors occur in approximately 10% of the patients. Immune deficiencies occur in both T- and B-cell limbs and are quite variable even within the same patient. Low IgA levels are detected in 80% of the patients studied, and IgD levels are lower, while IgG is usually normal. Monomeric IgM has been found in 80% of the patients. T-Cell immunity is impaired, as there is hypoplasia of the thymus, usually without Hassall's corpuscles. The peripheral and central lymphoid organs are characteristically depleted. Although the underlying pathogenetic mechanism is not known, the thymic abnormality hardly can account for the multisystemic nature of the disease. Recently discovered defects in DNA repair mechanisms in this disorder may be the cause of the high frequency of spontaneous chromosomal breaks, the sensitivity to irradiation, and the development of neoplasia, as well as the defects in CMI that plague ataxia telangiectasia patients, and also the high incidence of lymphoreticular malignancy.

Defects of Regulation

Increased suppressor T-cell activity has been demonstrated in several patients with common variable hypogammaglobulinemia or with isolated Ig deficiencies. Ig production suppression has also been observed in patients with hypogammaglobulinemia and thymoma. Evidence for increased suppressor T-cell activity has been obtained from coculture experiments (Ig synthesis by normal B cells plus helper T cells) and from demonstrations that *in vivo* Ig levels became normal after corticosteroid treatment (these drugs selectively affect suppressor T cells).

Finally, a group of patients, originally represented from a single kindred, now known in several families, have been shown to have fatal infectious mononucleosis, aproliferative disorders (agammaglobulinemia), and lymphoproliferative disease. This syndrome, designated X-linked recessive lymphoproliferative syndrome (XLRLS) or Duncan's syndrome, seems to be related to aberrant lymphocyte populations and, perhaps, a defect in natural killer cells. The common denominator appears to be abnormal lymphocyte responses following Epstein–Barr (EB) viral infections.

GENETICS OF COMPLEMENT DEFICIENCY

Genetic defects in humans may affect C1r, C4, C2, C3, C5, C6, C7, and C8. These deficiencies are inherited as autosomally recessive traits. Also, a deficiency of $C\bar{1}$ inhibitor, which leads to hereditary angioneurotic edema, is inherited as an autosomal dominant. The defect in C3b inactivator has been reported to be inherited as an autosomal recessive. Table 7.4 contains a list of some of the well-recognized complement deficiencies. Most defects are inherited as autosomally recessive traits; thus the heterozygotes are detected easily, because their sera contain around half the normal levels of the complement component. The genes for C2 and C4 map within the major HLA complex (see Chapter 5). In several cases of C2 deficiency, the HLA haplotype *HLA-A10, B18*, to which the C2 gene is linked, were also found. This is an extraordinary incidence compared with that in the general population. C2, C4, Factor B, and the receptors for C3b and C3d are all coded for on chromosome 6 (as is HLA). The comple-

Table 7.4 Complement deficiencies (Reprinted from WHO Report on Immunodeficiencies from *Clin. Immunol. Immunopathol.* 13: 296, 1979)

Deficiency	Inheritance*	HLA linkage	Symptom
C1r	AR	—	SLE-like syndrome
C4	AR	+	SLE-like syndrome
C2	AR	+	SLE-like syndrome: vasculitis; polymyositis, arthritis
C3	AR	—	Recurrent pyogenic infections
C5	AR	—	SLE
C6	AR	?	Neisserial infection
C7	AR	—	Neisserial infection
C8	AR	—	Neisserial infection
C1 inhibitor	AD	—	Hereditary angioedema
C3b inactivator	AR	—	Recurrent pyogenic infection

* AR = autosomal recessive; AD = autosomal dominant.

ment components are very polymorphic, and the association between different allotypes and disease is an important area of study.

Patients with hereditary angioneurotic edema and those with C4 or C2 deficiencies (or any isolated complement component) suffer from systemic lupus erythematosus (SLE) and glomerulonephritis far more often than the normal population. For the most part, isolated complement deficiencies seem to predispose to recurrent bacterial infections. When the patient has SLE, recurrent infections but normal immunoglobulins, or severe neisserial infections, it should raise the suspicion of an inherited complement component deficiency.

IMMUNODEFICIENCY STATES IN ANIMALS

In the mouse, the presence of the *Fv-1* locus on the fourth chromosome determines whether leukemia viruses (MuLV) can replicate. Susceptibility to Friend virus-induced leukemia is influenced by several genes (S1, *W*, and *f*) that affect host target cell metabolism. Although unigenic control of certain murine viral infections has been demonstrated, polygenic control thus far is the rule. Also, although some viral immunity in experimental animals has been shown to be under dominant genetic control (e.g., mouse pox and lymphocytic choriomeningitis), most resistance is seen to be under polygenic control, with some degree of histocompatibility involvement. Genetic resistance to C-type viruses appears to be under polygenic control.

Recently, genes that influence viral oncogenesis have been discovered in inbred mouse lines. Highly susceptible (or resistant) lines have been developed, and studies have left little doubt that neoplasia is under genetic control. Chang and Hilderman identified three independently segregating loci that account for susceptibility of mice to polyoma virus. Interestingly, susceptibility to actual tumorigenesis seemed to be determined by a single autosomal gene with incomplete dominance. *In vitro,* the cells were uniformly susceptible to viral infection and replication. The decisive component in genetic resistance *in vivo* seemed to be the early maturation of the immune system.

Certain loci exist that control virus-induced tumorigenesis (MuLV) by affecting the expression of viral immunogens or by regulating the immune response to virus or virally transformed cells. The resistance determined by these genes is inherited dominantly. Two independent loci have been implicated in determining resistance to Gross leukemia virus: *Rgv-1* is closely linked to H-2; the location of *Rgv-2* is unknown. Studies indicate that *Rgv-1* maps to the *K* end of the H-2 complex. Also, in the *K* end of H-2 is *Rfv-2* and *Rrv-1* in the *I* region, which control resistance to Friend and radiation leukemia virus, respectively. These loci may be *Ir* genes analogous to the control by H-2 linked genes of antibodies to Gross virus-specific antigens and the cell-mediated cytotoxicity toward AKR leukemia that is under *I*-region control.

At least six autosomal recessive mutations thus far have been discerned in mice that lead to T-cell dysfunction. The most widely examined is the nude mouse (*nu/nu*), which lacks a thymus and hair. This experimental model has been very helpful in studying T-cell-dependent immunity. Nude mice fail to demonstrate allograft or even xenograft rejection, to make antibody to the T-dependent antigens, or to respond to mitogens that normally activate T cells. Although nude mice fail to maintain normal levels of IgA (and sometimes IgG), serum IgM levels are normal. Syngeneic thymus grafts correct the immune disorder. Thus nude mouse bone marrow cells are normal, and some can be induced to express T-cell differentiation antigens on exposure to thymic hormones (thymopoietin).

Nude mice have been employed to scrutinize the role of T-cell immunological surveillance in oncogenesis. Some tumors do not grow well or at all in nude mice. Also, nude mice seem to be as resistant, if not more so, to chemically induced tumors. This seems to be due to normal levels of natural killer (NK) cells in the nude mouse. These findings have led recently to an intense effort directed at identifying NK cells and exploiting their mechanism of tumor surveillance.

Mortality rates for neoplasia in primary immunodeficiency are 100–200 times that expected in the general population. Leukemias, malignant lymphomas, or reticulum cell sarcomas seem to be the most frequent types of cancers associated with immunodeficiency. Approximately 10% of those individuals with ataxia telangiectasia, Wiscott–Aldrich syndrome, Bloom's syndrome, and X-linked agammaglobulinemia eventually suffer from cancer. Reasons for these associations may include (1) the classical immune surveillance theory, (2) chronic stimulation by antigen in immunodeficient patients leading to mutation and transformation, and (3) a loss of normal suppressor T cells, leading to chronic antigenic stimulation, mutation, and neoplastic transformation.

Investigations concentrating on the study and manipulation of several spontaneously occurring animal models with genetically associated immune defects have furthered our understanding of immune deficiency and/or hyperactivity in the so-called autoimmune diseases.

In 1977, two families of chickens were discovered that showed low 7s Ig production but greatly elevated amounts of serum IgM (Benedict *et al.*, 1978). These chickens, known as UCD140, present overt disease that resemble certain forms of human variable immunodeficiency. Bursal cells are markedly underdeveloped, and in coculture T cells selectively suppress 7s Ig but not IgM biosynthesis. Also, pokeweed mitogen (PWM) stimulates their IgM but not IgG production. The antibody response to T-dependent antigens is especially deficient in IgG production. These birds develop an autoimmune constellation at 6 months including immune complex disease, Coombs'-positive hemolytic anemia, cryoglobulins, and serum rheumatoid factor. Interestingly, UCD140 B cells were able to synthesize 7s IgG when supplied with histocompatibly normal T cells. The genetic basis of this disease does not appear to be associated with the major histocompatibility complex; however, genetic backcross analysis suggests that genetic control is by more than one gene with additive action but variable penetrance.

An animal model for severe combined immune deficiency in children exists in the Arabian foal. These animals lack IgM and any other immunoglobulin class at birth. This defect is inherited as an autosomal recessive. The thymus is extremely hypoplastic, and peripheral blood lymphocytes fail to respond to both B- and T-cell-specific mitogens. The mode of inheritance in this disorder is unknown (Gershwin and Merchant, 1981).

Takatsu and Hamaoka (1982) have discovered that the DBA/2Ha inbred strain of mouse demonstrates an X-linked immunodeficiency involving its B lymphocytes. Thus the B cells of DBA/2Ha mice are selectively deficient in membrane receptors for T-cell replacing factor (TRF) produced by T cells, which prevents the triggering of the B cell by intravenous injections of anti-TRF. This murine model of an immunodeficiency disease may provide the strategy necessary for physiochemical characterization of such TRF acceptor sites and for distinguishing this site from other B-cell components such as Ia antigen, Fc receptor, and Lyb markers. This contrasts with another X-linked immune deficiency in CBA/N mice, which are unable to respond to certain types of antigens that normally trigger IgM and IgG3 antibodies. Mice that express the CBA/N X^{id} gene lack Lyb 3, 5, 7 and IaW39 membrane-positive B cells. This immunodeficiency model has helped document the existence of distinct subpopulations of B cells (Quintans, 1977).

Another animal model, whose similarity to systemic lupus erythematosus of humans has fostered intense research, is the Aleutian disease of mink, caused by a temperature-sensitive, nondefective, xenotropic parvovirus (ADV); these animals die at from 2 months to 2 years of age from an immune-complex-mediated kidney lesion (Gershwin and Merchant, 1981). Enhancement of B-cell responses to ADV is marked in these animals, but whether this results from diminished immunosuppression or inherent enhanced B-cell responsiveness or decreased "B-cell responsiveness to suppression" is not known.

Inbred mice with autoimmune disease have provided probably the most important models for the genetic analysis of this set of immune disturbances. The oldest and best studied model (last 20 years) involves the New Zealand black and New Zealand white hybrid (NZB/NZW). These mice show a lupuslike autoimmune disease with polyclonally activated B cells, T-suppressor cells, natural thymocytotoxic antibodies, and a refractoriness to T-cell tolerance induction. A large number of genetic studies have shown that their disease is under polygenic control.

The genetic defect that leads to autoimmunity in NZB mice seems to be expressed in the bone marrow, since adoptively transferred NZB bone marrow will cause disease in normal irradiated hosts. NZB mice seem to have a suppressor T-cell deficiency in early life, and these cells may be resistant to tolerance induction. NZB mice at 6 months have excessive numbers of Lyt1 + T cells and too few Lyt123 + T cells. This latter cell subset appears to be responsible for suppression. It has been suggested that natural thymocytotoxic antibodies are better in killing the Lyt123 + cell (suppressor) than the helper T cells. By 3 months of age, NZB mice show a high incidence of chromosomal anomalies including euploidy and aneuploidy. Whether these cause or result from the autoimmune process is unclear. The chromosome that houses the *H-2* gene locus (number 17) is found to be hyperploid, which could profoundly affect a number of *Ir* genes. Trisomy 15 of NZB spleen cells is also found in many spontaneous, radiation- and virus-induced leukemias, T-cell lymphomas, and reticulum cell sarcomas. Other murine models, investigations of which have shed much light on autoimmune phenomena, are BXSB and MRL/1 (Dixon, 1982). In the BXSB strain, evidence indicates that the lupus is under polygenic control, is specifically associated with the Y chromosome, and is not hormonally mediated. This lack of hormonal control contrasts with the situation in NZB/NZW mice, where female sex hormones can accelerate the disease. The BXSB strain demonstrates spontaneous autoimmune disease with Coombs'-positive hemolytic anemia, nephritis, and hypergammaglobulinemia.

The MRL mice demonstrate 90% mortality by 7 months in females and 9 months in males. They have an early-onset autoimmune disease with concomitant autoantibodies, nephritis, and a massive lymphadenopathy with proliferation of Lyt1-positive T cells and Thy-1-positive null cells. The MRL/1 genome contains the lymphoproliferative (*lpr*) gene that is recessive and when homozygous produces the massive lymphoid proliferation noted above. Pisetsky *et al.* (1982) has developed by mating a strain, B6lpr/lpr, that would allow evaluation of this *lpr* gene on a completely innocuous background. They discovered that this *lpr* gene is responsible for controlling, and not merely accelerating a predisposition for, the anti-DNA antibody response. Thus the lupus-accelerating factors enumerated by Dixon (1982) include (1) the *lpr* gene in MRL mice, (2) genes on the Y chromosome in BXSB mice, (3) female hormones in New Zealand mice, and (4) chronic lymphochoriomeningitis virus (LCM) infection at birth leading to immune-complex-mediated nephrology.

Although there have been numerous reports implicating suppressor T-cell defects in lupuslike etiology, recent evidence (Theofilopoulos *et al.*, 1982; Dixon, 1982) consistently and repeatedly shows that one of the most characteristic abnormalities of all lupus animals is a B-cell hyperactivity. The B-cell system may mature earlier than normal, as reflected in an earlier IgM-to-IgG switch and in early appearance of surface membrane receptors. Thus, there may be an increased IgG : IgM receptor ratio and premature expression of C3 receptors.

Human Chediak–Higashi syndrome has counterparts in several animal models, including cow, cat, mink, killer whale, and mouse. The beige murine shows defective granulocyte chemotaxis, elevated serum IgM, and defective natural killer cells. Recent evidence demonstrates that these defects may be due to aberrant cyclic nucleotide metabolism and microtubule polymerization (Gershwin and Merchant, 1981).

In terms of genetically determined organ-specific autoimmune disease, obsese-strain (OS) chickens suffer from a hypothyroidism due to an early and severe immunologically mediated destruction of the thyroid gland. This strain is not subject to development of other autoimmune diseases, but defects in the thyroid gland itself may predispose to a thyroiditis with complement-fixing antibodies directed toward thyroglobulin. Several types of evidence now incriminate a hyperactive peripheral B-cell system in OS chickens (Gershwin and Merchant, 1981). An elevated B : lymphocyte ratio, elevated antibody titers to erythrocyte nuclei, and significantly high titers of natural heteroagglutinins combine to indict the B cell. The bursal system appears to be intact and functions normally. In this disease, it appears that alleles within the chicken major histocompatibility complex (B complex) have a significant effect on this autoimmune thyroiditis. The relationships between these described cases of immunodeficiency, chromosomal aberrations, autoimmunity, and malignancy are currently areas of intense scientific investigation.

RECOMMENDED READINGS

Books

Asherson, G. L. and Webster, A. D. B. (1980), *Diagnosis and Treatment of Immunodeficiency Diseases*. Oxford and London: Blackwell.

Hildermann, W. H., Clark, E. A., and Rason, R. L. (1981), *Comprehensive Immunogenetics*. New York: Elsevier North-Holland.

Rose, N. R., Biagazzi, P. E., and Warner, N. L. (Eds.) (1978), *Genetic Control of Autoimmune Diseases*. New York: Elsevier North-Holland.

Cooper, M. and Lawton, A. R. (Eds.) (1978), *Immune Deficiency I and II*, Springer Seminars in Immunopathology (Vol. I). Berlin: Springer-Verlag.

Gershwin, M. E. and Merchant, B. (Eds.) (1981), *Immunologic Defects in Laboratory Animals* (Vols. I and II). New York: Plenum Press.

Klein, J. (1982), *Immunology, The Science of Self–Nonself Discrimination*. New York: Wiley, pp. 156–255.

Articles

Aiuti, F., Fontana, L., and Gatti, R. (1973), Membrane-bound immunoglobulin (Ig) and *in vitro* production of Ig by lymphoid cells from patients with primary immunodeficiencies. *Scand. J. Immunol.* 2, 9.

Benedict, A. A., Chanh, T. C., Tam, L. Q., Polland, L. W., Kubo, R. T., and Abplanalo, H. A. (1978), Immunodeficiency in chickens: a model for common variable hypogammaglobulinemia in men? *Adv. Exp. Med. Brit.* 88, 197.

Berkel, A. I., Caglar, M., Tinatepe, K., and Ersoy, F. (1976), Thymic dysplasia with normal immunoglobulins (the Nezelof syndrome). (A case report). *Turk. J. Pediatr.* 18, 45.

Carson, D. A. and Seegmiller, J. E. (1976), Effect of adenosine deaminase inhibition upon human lymphocyte blastogenesis. *J. Clin. Invest.* 57, 274.

Chang, S. S. and Hildermann, W. H. (1968), At least three independently segregating loci account for overall control of susceptibility to polyoma virus. *J. Natl. Cancer Inst.* 40, 363.

Chang, S. S. and Hildermann, W. H. (1964), Inheritance of susceptibility to polyoma virus in mice. *J. Natl. Cancer Inst.* 33, 303.

Cooper, M. D. and Lawton, M. R. (1972), Circulating B cells in patients with immunodeficiency. *Am. J. Pathol.* 69, 513.

Davis, S. D., Schaller, J., and Wedgwood, R. J. (1966), Job's syndrome: recurrent "cold" staphylococcal abscesses. *Lancet* 1, 1013.

DiGeorge, A. M. (1968), Congenital absence of the thymus and its immunologic consequences: concurrence with congenital hypoparathyroidism. In D. Bergsma (Ed.), *Immunologic Deficiency Diseases in Man.* New York: The National Foundation, March of Dimes.

Dissing, J. and Knudsen, B. (1972), Adenosine deaminase deficiency and combined immunodeficiency syndrome. *Lancet* 2, 1316.

Dixon, F. J. (1982), Murine SLE models and autoimmune disease. *Hosp. Pract.* 17, 63–73.

Donabedian, H. and Gallin, J. I. (1982), Mononuclear cells from patients with the hyperimmunoglobulin E-recurrent infection syndrome produce an inhibition of leukocyte chemotaxis. *J. Clin. Invest.* 69, 1155.

Douglas, S. D. and Fudenberg, H. H. (1969), Graft-versus-host reaction in Wiskott-Aldrich syndrome: antemortem diagnosis of human GVH in an immunologic deficiency disease. *Vox Sang.* 16, 172.

Faulk, W. P., Kiyasu, W. S., Cooper, M. D., and Fudenberg, H. H. (1971), Deficiency of IgM. *Pediatrics* 47, 399.

Fialkow, P. J., Bryant, J. I., and Friedman, J. M. (1978), Specificity of acquired clonal chromosome abnormalities in New Zealand black mice. *Int. J. Cancer* 21, 505.

Fudenberg, H. H., German, J. L., and Kunkel, H. G. (1962), The occurrence of rheumatoid factor and other abnormalities in families of patients with agammaglobulinemia. *Arthritis Rheum.* 5, 565.

Fudenberg, H. and Solomon, A. (1961), Acquired agammaglobulinemia with autoimmune hemolytic disease: graft-versus-host reaction. *Vox Sang.* 6, 68.

Gatti, R. A. (1974), On the classification of patients with primary immunodeficiency disorders. *Clin. Immunol. Immunopathol.* 3, 243.

Good, R. A., Kelly, W. D., Rotstein, J., and Varcey, R. L. (1962), Immunological deficiency diseases, agammaglobulinemia, hypogammaglobulinemia, Hodgkin's disease and sarcoidosis. *Prog. Allergy* 6, 181.

Good, R. A., Rotstein, J., and Mazzitello, W. F. (1957), The simultaneous occurrence of rheumatoid arthritis and agammaglobulinemia. *J. Lab. Clin. Med.* 48, 343.

Gupta, S. and Good, R. A. (1977), Subpopulations of human T lymphocytes. I. Studies in immunodeficient patients. *Clin. Exp. Immunol.* 30, 222.

Higashi, O. (1954), Congenital gigantism of the peroxidase granules. The first case ever reported of qualitative abnormality of peroxidase. *Tohoku J. Exp. Med.* 59, 315.

Jersild, C., Rubinstein, P., and Day, N. K. (1976), The HLA system and inherited deficiencies of the complement system. *Transplant. Rev.* 32, 43.

Jeunet, F. and Good, R. A. (1968), Thymoma, immunologic deficiencies and hematological abnormalities. In D. Bergsma (Ed.), *Immunologic Deficiency Diseases in Man.* New York: The National Foundation, March of Dimes.

Litwin, S. D. and Fudenberg, H. H. (1972), Quantitative abnormalities of allotypic genes in families with primary immunodeficiencies. *Proc. Natl. Acad. Sci. (USA)* 69, 1739.

Malawista, S. E. (1978), Simple screening test for chronic granulomatous disease of childhood. *Lancet* 1, 943.

Najjar, V. A. (1975), Defective phagocytosis due to deficiencies involving the tetrapeptide tuftsin. *J. Pediatr.* 87, 1121.

O'Reilly, R. J., Pahwa, R., Dupont, B., and Good, R. A. (1978), Severe combined immunodeficiency: transplantation approaches for patients lacking an HLA genotypically identical sibling. *Transplant. Proc.* 10, 187.

Oxelius, V. A. (1974), Chronic infections in a family with hereditary deficiency of IgG2 and IgG4. *Clin. Exp. Immunol.* 17, 19.

Pandey, J. P., Fudenberg, H. H., Virella, G., Kyong, C. U., Loadholt, C. B., Galbraith, R. M., Gotschlich, E. C., and Parke, J. C. (1979), Association between immunoglobulin allotypes and immune responses to *Haemophilus influenzae* and meningococcal polysaccharides. *Lancet* 1, 190.

Peterson, R. D., Cooper, M. D., and Good, R. A. (1966), Lymphoid tissue abnormalities associated with ataxia telangiectasia. *Am. J. Med.* 41, 342.

Pincus, T. (1982), Studies regarding a possible function for viruses in the pathogenesis of systemic lupus erythematosus. *Arthritis Rheum.* 25, 847.

Pisetsky, D. S., Caster, S. A., Roths, J. B., and Murphy, E. D. (1982), *Ipr* gene control of the anti-DNA antibody response. *J. Immunol.* 128, 2322.

Purtilo, D. T. (1976), Pathogenesis and phenotypes of an X-linked recessive lymphoproliferative syndrome. *Lancet* 2, 882.

Purtilo, D. T., Szymanski, I., Bhawan, J., Yang, J. P. S., Hutt, L. M., Boto, W., DeNicola, L., Maier, R., and Thorley-Lawson, D. (1978), Epstein–Barr virus infections in the X-linked recessive lymphoproliferative syndrome. *Lancet* 1, 798.

Quintans, J. (1977), The "patchy" immunodeficiency of CBA/N mice. *Eur. J. Immunol.* 7, 749.

Ross, I. N. and Thompson, R. A. (1976), Severe selective IgM deficiency. *J. Clin. Pathol.* 29, 773.

Rowe, W. P. (1973), Genetic factors in the natural history of murine leukemia virus infection. *Cancer Res.* 33, 3061.

Segal, A. W., Cross, A. R., Garcia, R. C., Borregaard, N., Valerius, N. H., Soothill, J. F., and Jones, O. T. G. (1983), Absence of cytochrome b-245 in chronic granulomatous disease *N. Engl. J. Med.* 308, 245.

Siegel, F. P. and Good, R. A. (1977), Human and lymphocyte differentiation markers and their applications to immunodeficiency and lymphoproliferative diseases. *Clin. Hematol.* 6, 355.

Soothill, J. F. and Harver, B. A. M. (1976), Defective opsonization: a common immune deficiency. *Arch. Dis. Child.* 51, 91.

Spitler, L. E., Levin, A., Stites, D., Fudenberg, H., and Huber, H. H. (1975), The Wiskott–Aldrich syndrome. Immunologic studies in nine patients and selected family members. *Cell. Immunol.* 19, 201.

Takatsu, K. and Hamaoka, T. (1982), DBA/2Ha mice as a model of an X-linked immunodeficiency which is defective in the expression of TRF-acceptor site(s) on B lymphocytes. In G. Möller (Ed.), *Immunological Reviews* (Vol. 64). Copenhagen: Munksgaard, pp. 25–55.

Theofilopoulos, A. N., Balderes, R. S., Gozes, Y., Fidler, J. M., Liu, F. T., Ahmead, A., and Dixon, F. J. (1982), Surface and functional characteristics of B cells from lupus-prone murine models. *Clin. Immunol. Immunopathol.* 23, 224.

Tominaga, A., Takatsu, K., and Hamaoka, T. (1980), Antigen-induced T cell-replacing factor. II. X-linked gene control for the expression of TRF acceptor site(s) on B lymphocytes and preparation of specific antiserum to that acceptor. *J. Immunol.* 124, 2423.

Tzehoval, E., Segal, S., Stabinsky, Y., Fridkin, M., Spirer, Z., and Feldman, M. (1978), Tuftsin (an Ig-associated tetrapeptide) triggers the immunogenic function of macrophages: implications for activation of programmed cells. *Proc. Natl. Acad. Sci.* (*USA*) 75, 3400.

Uytde Haag, F., Heynen, C. J., and Ballireux, R. E. (1978), Induction of antigen-specific human suppressor T lymphocytes *in vitro*. *Nature* 271, 556.

Webster, A. D. B. (1980), Infections in immunodeficient patients. In D. A. J. Tyrrell (Ed.), *The Biology of Some Slow and Persistent Virus Infections*. The Hague: Martinus Nijhoff.

WHO Scientific Group on Immunodeficiency (1979), *Clin. Immunol. Immuno-Pathol.* 13, 296.

APPENDIX 1

Glossary of Immunological Terms

Accessory Cells: Lymphoid cells of the monocyte and macrophage lineage that cooperate with T and B lymphocytes in the formation of antibody and in other immune reactions.

"Activated" Lymphocytes: Lymphocytes that have been stimulated by specific antigen or nonspecific mitogen.

Adenosine Deaminase: An enzyme that catalyzes the conversion of adenosine to inosine and is deficient in some patients with combined immunodeficiency syndrome.

Affinity: A measure of attraction between antigen and antibody that refers to closeness of stereochemical fit (binding).

Allele: One of two genes controlling a particular characteristic present at a locus.

Allelic Exclusion: The phenotypic expression of a single allele in cells containing two different alleles for that genetic locus.

Allogeneic: Denotes the relationship that exists between genetically dissimilar members of the same species.

Allogeneic Effect: A form of general immunopotentiation in which specific stimulation of T cells results in the release of factors active in the immune response.

Allograft (also called homograft): A tissue or organ graft between two genetically dissimilar members of the same species.

Allotype: The genetically determined antigenic difference on serum proteins, varying in different members of the same species.

Alternative Complement Pathway (also known as the properdin pathway): The system of activation of the complement path-

Adapted in part from H. H. Fudenberg, D. P. Stites, J. V. Wells, and J. Caldwell (1976), *Basic and Clinical Immunology*. Los Altos, Calif.: Lange Medical Publications, pp. 615–623, by permission of the publisher.

way through involvement of properdin factor D, properdin factor B, and C3b, finally activating C3 and then progressing as in the classical pathway.

Am marker: Allotypic determinant on the H chain of human IgA.

Anergy: The inability to react to a battery of common skin test antigens.

Antibody: A protein that is produced as a result of the introduction of an antigen and that has the ability to combine with the antigen that stimulated its production.

Antibody-Combining Site: The configuration present on an antibody molecule that links with a corresponding antigenic determinant.

Antibody-Dependent Cell-Mediated Cytotoxicity (ADCC): A form of lymphocyte-mediated cytotoxicity in which an effector cell kills an antibody-coated target cell, presumably by recognition of the Fc region of the cell-bound antibody through an Fc receptor present on the effector lymphocyte.

Antigen: A substance that can induce a detectable immune response when introduced into an animal.

Antigen-Binding Site: The part of an immunoglobulin that binds antigen.

Antigenic Determinant (*also called epitope*): The area of an antigen that determines the specificity of the antigen–antibody reaction.

Antigen Processing: The series of events that occurs following antigen administration and antibody production.

Antiglobulin Test (*Coombs' test*): A technique for detecting cell-bound immunoglobulin. In the direct Coombs' test, red blood cells taken directly from a sensitized individual are agglutinated by anti-gammaglobulin antibodies. In the indirect Coombs' test, a patient's serum is incubated with test red blood cells and the sensitized cells are then agglutinated with an anti-immunoglobulin or with Coombs' reagent.

Antilymphocyte Serum: Antibodies that are directed against lymphocytes (and that usually causes immunosuppression).

"Armed" Macrophages: Macrophages capable of antigen-specific cytotoxicity as a result of cytophilic antibodies or arming factors from T cells.

Atopy: A genetically determined abnormal state of hypersensitivity as distinguished from hypersensitivity responses in normal individuals, which are not genetically determined.

Attenuated: Rendered less virulent.

Autoantibody: Antibody to self antigens.

Autoantigens: Self antigens.

Autoimmunity: Immunity to self antigens (autoantigens).

Autoradiography: A technique for detecting radioactive isotopes in which a tissue section containing radioactivity is overlaid with X-ray or photographic film on which the emissions are recorded.

Avidity: The quantitative measurement for affinity (usually in Kcal. $mole^{-1}$); sometimes used synonymously with affinity.

B cell (also known as B lymphocyte): Strictly, a bursa-derived cell in avian species; by analogy, bursa-equivalent derived cells (also called "bone-marrow-derived") in nonavian species. B cells are the precursors of the plasma cells that produce antibody.

Bence–Jones Proteins: Monoclonal L chains present in the urine of patients with paraproteinemic disorders.

Blocking Factors: Substances present in the serum of tumor-bearing animals that block the ability of immune lymphocytes to kill tumor cells.

C Region (constant region): The carboxy-terminal portion of the H or L chain that is identical in immunoglobulin molecules of a given class and subclass (apart from genetic polymorphisms).

C Terminal: The carboxy-terminal end of a protein molecule.

Capping: The movement of cell-surface antigens toward one pole of a cell after the antigens are cross-linked by specific antibody.

Carrier: An immunogenic substance that, when coupled to a hapten, renders the hapten immunogenic.

Cationic Proteins: Antimicrobial substances present within granules of phagocytic cells.

Cell-Mediated Immunity: Immunity in which the participation of T lymphocytes and macrophages is predominant.

Cell-Mediated Lymphocytolysis: An *in vitro* assay for cellular immunity in which a standard mixed lymphocyte reaction is followed by destruction of target cells which are used to sensitize allogeneic cells during the MLC.

Central Lymphoid Organs: Lymphoid organs that are essential to the development of the immune response, i.e., the thymus and the bursa of Fabricius.

Chemotaxis: A process whereby phagocytic cells are attracted to the vicinity of invading pathogens.

Chromatography: A variety of techniques useful for the separation of proteins.

Classical Complement Pathway: A series of enzyme–substrate and protein–protein interactions that ultimately leads to biologically active complement enzymes.

Clonal Selection Theory: The theory of antibody synthesis proposed by Burnet, which predicts that the individual carries a complement of clones of lymphoid cells that are capable of reacting with all possible antigenic determinants. During fetal life, clones reacted against self antigens are eliminated on contact with antigens.

Clone: A group of cells all of which are the progeny of a single cell.

Codominant: An equal expression of the two alleles in a pair of genes in the heterozygous state.

Cohn Fraction II: Primarily gammaglobulin that is produced as the result of ethanol fractionation of serum according to the Cohn method.

Cold Agglutinins: Antibodies that agglutinate bacteria or erythrocytes more efficiently at temperatures below 37°C than at 37°C, and optimally at 4°C.

Complement: A cascade system of serum proteins that is the primary amplification mechanism of antigen–antibody reactions.

Complement Fixation: A standard serological assay used for the detection of an antigen–antibody reaction in which complement is fixed as a result of the formation of an immune complex. The subsequent failure of lysis of sensitized red blood cells by complement that has been fixed indicates the degree of antigen–antibody reaction.

Complotype: A haplotype composed of closely linked alleles coding for complement components.

Concanavalin A (ConA): A lectin that is derived from the jack bean and that stimulates predominantly T lymphocytes.

Congenic (originally called congenic resistant): Denotes a line of mice identical or nearly identical with other inbred strains except for the substitution at one histocompatibility locus of a foreign allele introduced by appropriate crosses with a second inbred strain.

Constant Region: The carboxy-terminal portion of immunoglobulins. Within classes, it remains relatively constant in amino acid content and arrangement from molecule to molecule, except for allotypic differences.

Contact sensitivity: A type of delayed hypersensitivity reaction in which sensitivity to simple chemical compounds is manifested by skin reactivity.

Cryoglobulin: A protein that has the property of forming a precipitate or gel in the cold.

Cytotoxicity: The lysing of cells, which is accomplished by antibody and complement, antibody and cells (ADCC), T cells, NK cells, or macrophages.

Degranulation: A process whereby cytoplasmic granules of phagocytic cells fuse with phagosomes and discharge their contents into the phagolysosome thus formed.

Delayed Cutaneous Hypersensitivity: A cell-mediated immune reaction that can be elicited by subcutaneous injection of antigen, with a subsequent cellular infiltrate and edema that are maximal between 24 and 48 hours after antigen challenge.

Dextrans: Polysaccharides composed of a single sugar.

Direct Agglutination: Agglutination of red cells, microorganisms, or other substances directly by serum antibody.

Disulfide Bonds: Chemical S—S bonds between sulfhydryl-containing amino acids, which bind together H and L chains as well as portions of H—H and L—L chains.

DNA Cloning Technique: The process of inserting a particular section of DNA (gene) into a vehicle (bacterium) so that the gene is functional in the growing bacterium.

Domains (also called homology regions): Segments of H or L chains that are folded three-dimensionally and stabilized with disulfide bonds.

Dysgammaglobulinemia: Selective deficiency of one or more immunoglobulin classes (term no longer in general use).

E Rosette: Formation of a cluster (rosette) of cells consisting of sheep erythrocytes and human T lymphocytes.

EAC Rosette: Formation of a cluster of red cells (erythrocytes) sensitized with amboceptor (antibody) and complement around human B lymphocytes.

Effector Cells: Usually denotes T cells capable of mediating cytotoxicity, suppression, or helper function.

Epitope: The simplest form of an antigenic determinant present on a complex antigenic molecule.

Equilibrium Dialysis: A technique for measuring the strength or affinity with which antibody binds to antigen.

Equivalence: A ratio of antigen–antibody concentration where maximal precipitation occurs.

F_1 Generation: The first generation of offspring after a designated mating.

F_2 Generation: The second generation of offspring after a designated mating.

Fab: An antigen-binding fragment produced by enzymatic digestion of an IgG molecule with papain.

$F(ab')_2$: A fragment obtained by pepsin digestion of immunoglobulin molecules containing the two H and two L chains linked by disulfide bonds. It contains antigen-binding activity. An $F(ab')_2$ fragment and an Fc fragment comprise an entire monomeric immunoglobulin molecule.

Fc Fragment: A crystallizable fragment obtained by papain digestion of IgG molecules. The Fc fragment consists of the C-terminal half of two H chains linked by disulfide bonds. It contains no antigen-binding capability but determines important biological characteristics of the intact molecule.

Fc Receptor: A receptor present on various subclasses of lymphocytes for the Fc fragment of immunoglobulins, which is present on various subclasses of lymphocytes (mainly B cells) and also on certain tumor cell lines.

Fetal antigen: A type of tumor-associated antigen that is normally present on embryonic but not adult tissues and that is reexpressed during the neoplastic process.

Forbidden Clone Theory: The theory proposed to explain autoimmunity, which postulates that lymphocytes capable of self-sensitization and effector function are present in tolerant animals, since they were not eliminated during embryogenesis.

FOUR Locus (4 locus): The second human histocompatibility (HLA) locus to be described, now called locus B.

Framework Region: The portion of the variable (V) region that is relatively invariant as compared to the hypervariable regions.

Freund's Complete Adjuvant: An oil–water emulsion that contains killed mycobacteria and enhances immune responses when mixed in an emulsion with antigen.

Freund's Incomplete Adjuvant: Contains all the elements of Freund's complete adjuvant with the exception of killed mycobacteria.

Gammaglobulins: Serum proteins with gamma mobility on electrophoresis that comprise the majority of immunoglobulins and antibodies.

Genetic Switch Hypothesis: A hypothesis that postulates that there is a switch in the gene controlling H-chain synthesis in plasma cells during the development of an immune response, so that IgM H-chain synthesis switches to IgG H chain.

Germ Line Theory of Antibody Synthesis: Predicts that information for synthesis of all types of antibody exists in the genome and that specific receptors are preformed on immunocompetent cells.

Gm Marker: Allotypic determinant on the H chain of human IgG.

Graft-Versus-Host (GVH) Reaction: The clinical and pathological sequelae of the reactions of immunocompetent cells in a graft against the cells of the histoincompatible and immunodeficient recipient.

H-2 Locus: The major genetic histocompatibility region in the mouse.

Haplotype: That portion of the phenotype determined by closely linked genes of a single chromosome.

Hapten: A substance that is not immunogenic but can react with an antibody of appropriate specificity.

Hb_sAg: The coat or envelope of hepatitis B virus.

Heavy (H-) Chain Diseases: A heterogeneous group of paraprotein disorders characterized by the presence of monoclonal but incomplete H chains without L chains in serum or urine.

Helper T Cells: A subtype of T lymphocytes that cooperate with B cells in antibody formation.

Hemagglutination Inhibition: A technique for detecting small amounts of antigen in which homologous antigen inhibits the agglutination of red cells or other particles coated with antigen by specific antibody.

Hematopoietic System: All tissues responsible for production of the cellular elements of peripheral blood. This term usually excludes strictly lymphocytopoietic tissue such as lymph nodes.

Hemolysin: Antibody or other substances capable of lysing red blood cells.

Hinge Region: The area of the H chains in the C region between the first and second C-region domains. It is the site of enzymatic cleavage into $F(ab')_2$ and Fc fragments.

Histocompatible: Sharing transplantation antigens.

HLA (human leukocyte antigen) Locus: The major histocompatibility genetic region in humans.

Homologous Antigen: An antigen that induces an antibody and reacts specifically with it.

Humoral: Pertaining to molecules in solution in a body fluid, particularly antibody and complement.

Humoral Immunity: Antibody-mediated immunity.

Hybridization Technique: The fusion of a single antibody-producing cell with a tumor cell for the purpose of immortalization; this hybrid clone is then producing monoclonal antibody.

Hypervariable Regions: At least four regions of extreme variability that occur throughout the V region of H and L chains and that determine the antibody-combining site of an antibody molecule.

Hypogammaglobulinemia: Deficiency of all major classes of serum immunoglobulins.

Ia Antigens (I region-associated antigens): Antigens that are controlled by *Ir* genes and are present on various tissues.

Idiotype: A unique antigenic determinant present on homogeneous antibody or myeloma protein. The idiotype appears to represent the antigenicity of the antigen-binding site of an antibody and is therefore located in the V region.

IgA: Predominant immunoglobulin class present in secretions.

IgD: Predominant immunoglobulin class present on human B lymphocytes.

IgE: Reaginic antibody involved in immediate hypersensitivity reactions; elevated in parasitic infections.

IgG: Predominant immunoglobulin class present in human serum.

IgM: A pentameric immunoglobulin comprising approximately 10% of normal human serum immunoglobulins, with a molecular weight of 900,000 and a sedimentation coefficient of 19S.

8S IgM: A monomeric IgM consisting of one monomer of five tetrameric subunits.

IgT: An immunoglobulin theoretically present on the surface of T cells that serves as an antigen receptor.

Immune Complexes: Antigen–antibody complexes.

Immune Response Genes (Ir genes): Genes that control immune responses to specific antigens.

Immune Surveillance: A theory that holds that the immune system destroys tumor cells, which are constantly arising during the life of the individual.

Immunodominant: That part of an antigenic determinant that is dominant in binding with antibody.

Immunoelectrophoresis: A technique combining an initial electrophoretic separation of proteins followed by immunodiffusion with resultant precipitation arcs.

Immunogen: A substance that, when introduced into an animal, stimulates the immune response. The term denotes any substance that is capable of stimulating an immune response, in contrast to a substance that can only combine with antibody, i.e., an antigen.

Immunogenicity: Property of a substance making it capable of inducing a detectable immune response.

Immunoglobulin: A glycoprotein composed of H and L chains which functions as antibody. All antibodies are immunoglobulins, but it is not certain that all immunoglobulins have antibody function.

Immunoglobulin Class: A subdivision of immunoglobulin molecules based on unique antigenic determinants in the Fc region of the H chains. In humans there are five classes of immunoglobulins designated IgG, IgA, IgM, IgD, and IgE.

Immunoglobulin Subclass: A subdivision of the classes of immunoglobulins based on structural and antigenic differences in the H chains. For human IgG there are four subclasses: IgG1, IgG2, IgG3, and IgG4.

Indirect or Passive Agglutination: The agglutination of particles or red blood cells to which antigens have been coupled chemically.

Interferon: A heterogeneous group of low-molecular-weight proteins elaborated by infected host cells that protect noninfected cells from viral infections.

Inv Marker: See *Km marker.*

I Region: That portion of the major histocompatibility complex that contains genes that control immune responses.

Ir genes: See *Immune response genes.*

Isoagglutinin: An agglutinating antibody capable of agglutinating cells of other individuals of the same species in which it is found.

Isoallotypic Determinants: Genetic markers that act as allotypic determinants within a specific immunoglobulin subclass and also appear on all members of at least one other subclass of IgG.

Isoantibody: An antibody that is capable of reacting with an antigen derived from a member of the same species as that in which it is raised.

Isohemagglutinins: Antibodies to major red cell antigens present in members of a given species and directed against antigenic determinants on red cells from other members of the species.

J Chain: A polypeptide chain that is normally found in polymeric immunoglobulins, particularly IgA and IgM.

K-Cell: Killer cells (lymphoid or monocytoid) that mediate antibody-dependent cell-mediated cytotoxicity.

K and D Regions: Genetic loci in the major histocompatibility complex of the mouse, coding for H-2 antigens that are detectable serologically.

Kallikrein System: See *Kinin system.*

Kappa (κ) *Chains:* One of two major types of L chains.

Km Marker (*also called Inv*): Allotypic markers on the kappa L chain of human immunoglobulins.

LAF (*Interleukin l*): Lymphocyte-activating factor that is produced by macrophages especially when activated; induces the production of IL-2 from T cells.

LA *Locus:* The first human histocompatibility (HLA) locus to be described, now called locus A.

Lambda (λ) *Chain:* One of two major types of L chains.

Leukocyte Inhibitory Factor (*LIF*): A mediator of cellular immunity that inhibits the migration of polymorphonuclear leukocytes.

Light Chain (*L Chain*): Polypeptide chain present in all immunoglobulin molecules. Two types exist in most species and are termed kappa (κ) and lambda (λ).

Linkage Disequilibrium: An unexpected association of linked genes in a population.

Locus: The specific site of a gene on a chromosome.

Ly Antigens: Differentiation antigens present on thymocytes and peripheral T cells.

Lymphocyte Activation (*also known as lymphocyte stimulation, lymphocyte transformation, or blastogenesis*): An *in vitro* technique in which lymphocytes are stimulated to become metabolically active by antigen or mitogen.

Lymphocyte-Defined (*LD*) *Antigens:* A series of histocompatibility antigens that are present on the majority of mammalian cells and are detectable primarily by reactivity in the mixed lymphocyte reaction (MLR).

Lymphokines (*also termed "mediators of cellular immunity"*): Soluble products of lymphocytes (and in some cases other cell types) that are responsible for the multiple effects of a cellular immune reaction.

Lymphotoxin (*LT*): A lymphokine that results in direct cytolysis following its release from stimulated lymphocytes.

Macrophage Activation Factor (*MAF*): A lymphokine that will activate macrophages to become avid phagocytic cells.

Macrophage Chemotactic Factor: A mediator of cellular immunity that selectively attracts monocytes or macrophages to the area of its release.

Macrophages: Phagocytic mononuclear cells that derive from bone marrow monocytes and subserve accessory roles in cellular immunity.

Major Histocompatibility Complex (*MHC*): An as yet undetermined number of genes located in close proximity that determine histocompatibility antigens of members of a species.

Beta$_2$-Microglobulin: A protein (molecular weight 11,600) that is associated with the outer membrane of many cells, including lymphocytes, and that may function as a structural part of the histocompatibility antigens on cells.

Memory (*anamnestic response*): A secondary response that is powerful in that more antibody is produced or a quicker transplantation rejection is seen; due to existence of memory cells.

Migration Inhibitory Factor (*MIF*): A mediator of cellular immunity that is capable of inhibiting the migration of macrophages.

Mitogens (*also called phytomitogens*): Substances that cause DNA synthesis, blast transformation, and ultimately division of lymphocytes.

Mixed Lymphocyte Culture (*mixed leukocyte culture*) (*MLC*): An *in vitro* test for cellular immunity in which lymphocytes or leukocytes from genetically dissimilar individuals are mixed and mutually stimulate DNA synthesis.

Mixed Lymphocyte Reaction (*MLR*): See *Mixed lymphocyte culture.*

Monoclonal Hypergammaglobulinemia: An increase in immunoglobulins produced by a single clone of cells containing one H-chain class and one L-chain type.

Monoclonal Protein: A protein that is produced from the progeny of a single cell (called a "clone") and has only one H-chain and one L-chain type.

Monomer: The basic unit of an immunoglobulin molecule, consisting of four polypeptide chains (2 H and 2 L).

Morgan and Centimorgan: A unit of genetic map length; 1 morgan = 100 centimorgans; 1 centimorgan equals 1% crossing over between two loci.

Multiple Myeloma: A paraproteinemic disorder consisting typically of the presence of serum paraprotein, anemia, and lytic bone lesions.

Myeloma Protein: An intact monoclonal immunoglobulin molecule of any class other than IgM.

N Terminal: The amino-terminal end of a protein molecule.

Natural Killer Cell: Nonimmune cells (lymphocytes or monocytes) that kill foreign or virus-infected cells.

"Nonresponder": An animal that responds poorly to an antigen, usually because of genetic factors.

Nude Mouse: A hairless mouse that congenitally lacks a thymus and has a marked deficiency of thymus-derived lymphocytes.

Null Cells: Cells lacking the specific identifying surface markers for either T or B lymphocytes.

NZB Mouse: A genetically inbred strain of mice in which autoimmune disease resembling systemic lupus erythematosus develops spontaneously.

Oncogenesis: The process of producing neoplasia or malignancy.

Osteoclast Activating Factor (*OAF*): A mediator of cellular immunity that promotes the resorption of bone.

Paraproteinemia: A heterogeneous group of diseases characterized by the presence in serum or urine of a monoclonal immunoglobulin.

Peripheral Lymphoid Organs: Lymphoid organs not essential to the ontogeny of immune responses, i.e., the spleen, lymph nodes, tonsils, and Peyer's patches.

Phagocytes: Cells that are capable of ingesting particulate matter.

Phagocytosis: The engulfment of microorganisms or other particles by leukocytes.

Phagolysosome: A cellular organelle that is the product of the fusion of a phagosome and a lysosome.

Phylogeny: The developmental and evolutionary history of a group of animals.

Phytohemagglutinin (PHA): A lectin derived from the red kidney bean (*Phaseolus vulgaris*) that stimulates predominantly T lymphocytes.

Phytomitogens: Glycoproteins derived from plants that stimulate DNA synthesis and blast transformation in lymphocytes.

Plaque-Forming Cells: Antibody-producing cells capable of forming a hemolytic plaque in the presence of complement and antigenic erythrocytes.

Plasma Cells: Fully differentiated antibody-synthesizing cells that are derived from B lymphocytes.

Pokeweed Mitogen: A lectin derived from pokeweed (*Phytolacca americana*) that stimulates both B and T lymphocytes.

Polyclonal Hypergammaglobulinemia: An increase in gammaglobulin of various classes containing different H and L chains.

Polyclonal Mitogens: Mitogens that activate large subpopulations of lymphocytes.

Polyclonal Proteins: A group of molecules derived from multiple clones of cells.

Polymers: Immunoglobulins composed of more than a single basic monomeric unit (e.g., an IgA dimer consists of two units).

Precipitation: A reaction between a soluble antigen and soluble antibody in which a complex lattice of interlocking aggregates forms.

Private Determinant: A serologically defined alloantigen that is found only on one of several alleles.

Public Determinant: A serologically defined alloantigen found on two or more allelic products.

Pyroglobulins: Monoclonal immunoglobulins that precipitate irreversibly when heated to 56°C.

Radioimmunoassay: A variety of immunological techniques in which a radioactive isotope is used to detect antigens or antibodies in some form of immunoassay.

Recombinant: An animal that has experienced a recombinational event during meiosis, consisting of crossover and recombination of parts of two chromosomes.

Regulatory Gene: A gene that functions to turn on and off (i.e., regulate) other genes (structural genes).

Reticuloendothelial System: A mononuclear phagocytic system located primarily in the reticular connective tissue framework of the spleen, liver, and lymphoid tissues.

S Region: The chromosomal region in the H-2 complex containing the gene for a serum gammaglobulin.

S Value: Svedberg unit. Denotes the sedimentation coefficient of a protein, determined usually by analytic ultracentrifugation.

Second-Set Graft Rejection: Immunological rejection of a graft in a host that is immune to antigens contained in that graft.

Secretory IgA: A dimer of IgA molecules with a sedimentation coefficient of 11S, linked by J chain and secretory component.

Secretory Immune System: A distinct immune system that is common to external secretions and consists predominantly of IgA.

Secretory Piece (T Piece): A molecule of molecular weight 70,000 produced in epithelial cells and associated with secretory immunoglobulins, particularly IgA and IgM.

Sequential Determinants: Determinants whose specificity is dictated by the sequence of subunits within the determinant rather than by the molecular structure of the antigen molecule.

Serologically Defined (SD) Antigens: Antigens that are present on membranes of nearly all mammalian cells and are controlled by genes present in the major histocompatibility complex. They can easily be detected with antibodies.

Side-Chain Theory: Theory of antibody synthesis proposed by Ehrlich in 1896 suggesting that specific side chains that form antigen receptors are present on the surface membranes of antibody-producing cells.

Slow Virus: A virus that produces disease with a greatly delayed onset and protracted course.

Structural Gene: A gene whose DNA codes for the messenger RNA from which the amino acids are blueprinted into a polypeptide chain.

Suppressor T Cells: A subset of T lymphocytes that suppress antibody synthesis by B cells or inhibit other cellular immune reactions by effector T cells.

T Cell or T Lymphocyte: A thymus-derived cell that participates in a variety of cell-mediated immune reactions.

TCGF (Interleukin 2): T-Cell growth factor. It is produced by T cells and can be used to grow and clone T cells for long periods of time.

Theta Antigen: An alloantigen present on the surface of most thymocytes and peripheral T lymphocytes; now called Thy-1.

Thymus: The central lymphoid organ that is present in the thorax and controls the ontogeny of T lymphocytes.

Thymus-Dependent Antigen: Antigen that depends on T-cell interaction with B cells for antibody synthesis (e.g., erythrocytes, serum proteins, and hapten–carrier complexes).

Thymus-Independent Antigen: Antigens that induce an immune response without the apparent participation of T lymphocytes.

TL Antigen: A membrane antigen that is present on prothymocytes in mice with a TL + gene, but which is lost during thymic maturation.

Tolerance: Traditionally denotes that condition in which responsive cell clones have been eliminated or inactivated by prior contact with antigen, with the result that no immune response occurs on administration of antigen.

Transfer Factor (dialyzable leukocyte extract): a dialyzable extract of immune lymphocytes that is capable of transferring cell-mediated immunity in humans and possibly in other animal species.

Tryptic Peptides: Peptides produced as a result of tryptic digestion of a protein molecule.

Ultracentrifugation: High-speed centrifugation technique that can be used for the analytic identification of proteins with various sedimentation coefficients or as a preparative technique for separating proteins of different shapes and densities.

V (*variable*) *Region:* The amino-terminal portion of the H or L chain of an immunoglobulin molecule, containing considerable heterogeneity in the amino acid residues compared to the constant region.

Valence: The number of epitopes with which one antibody molecule can combine.

Viscosity: Physical property of serum that is determined by the size, shape, and deformability of serum molecules. The hydrostatic state, molecular charge, and temperature sensitivity of proteins.

Wasting Disease (also known as runt disease): A chronic, ultimately fatal illness associated with lymphoid atrophy in mice that are neonatally thymectomized.

Xenogeneic: Denotes the relationship that exists between members of genetically different species.

Xenograft: A tissue or organ graft between members of two distinct or different species.

Zone Electrophoresis: Electrophoresis performed on paper or cellulose acetate in which proteins are separated almost exclusively on the basis of charge.

APPENDIX 2

Blood Group Terminology

ABO System: One of the major human blood group systems discovered in 1900 by Landsteiner. The red cells carry the A, B, A and B, or neither antigen.

Absorption: The process by which an antibody is removed from a serum by allowing it to attach to red cells, carrying the appropriate antigen, that have been mixed with the serum. Once the antigen–antibody complexes have formed, the red cells (and attached antibody) are removed.

Adsorption: The uptake of antibody onto red cell surface antigens.

Allelic Gene or Allelomorph: An alternative form of a series of genes that can occupy a single locus on either of a pair of homologous chromosomes. Normally the gene products (expressed in terms of red cell antigens) are antithetical. As an example, Fy^a and Fy^b are a pair of *allelic genes,* and the antigens, $\mathrm{Fy^a}$ and $\mathrm{Fy^b}$, they cause to be produced are *antithetical* to each other.

Amorph: A gene that apparently fails to cause a specific antigenic determinant to be made. Note: some genes that at present are regarded as amorphs may in fact cause the production of determinants that cannot be detected serologically.

Antibody: A serum protein, produced as the result of the introduction of a foreign antigen, that has the ability to combine with (and in many instances destroy) the antigen that stimulated its production.

Antigen: A substance that, when introduced into the circulation of a subject lacking that antigen, can stimulate the production of a specific antibody.

Antigenic Determinant: That site on the antigen molecule that combines, specifically, with an antibody.

Bilirubin: A chemical produced as a breakdown product of destroyed red

* Reprinted from Peter D. Issitt, *Applied Blood Group Serology.* Oxnard, Calif.: Spectra Biologicals, 1970, pp. 153–154. By permission of the author.

blood cells. Normally present in small quantities in all sera following normal red cell destruction (at the end of the cell's life span), bilirubin is usually found present in abnormally high concentrations following increased *in vivo* red cell destruction.

Blocking Antibody: An antibody that has the ability to combine, specifically, with its antigen on the red cell surface but that is unable to cause agglutination in a saline system. Named because of its ability to engage all the receptor sites on the red cell surface and thus prevent agglutination of the cells if they are subsequently tested with a saline-agglutinating antibody of the same specificity.

Chromosome: The structures, located within the cell nucleus, that carry genes in a linear order. In humans, nucleated cells carry 23 pairs of chromosomes. At the time of mitosis, 23 like pairs of chromosomes are produced for the new cell. At meiosis the 23 pairs of chromosomes divide so that each sex cell formed carries one of each pair, or 23 single chromosomes.

Clone: A set of immunocytes (usually small lymphocytes or plasma cells) that are descendants of one common precursor cell.

"Coating" Antibody: See *"Incomplete" antibody.*

"Complete" Antibody: An antibody capable of causing visible agglutination of red cells in saline media.

Complement: A series of alpha and beta (mostly the latter) globulins found as normal constituents of virtually all sera. Some antigen–antibody reactions cause complement components to attach to the red cell surface.

Combining Site: Antigenic determinant.

Coombs' Serum: Anti-human-globulin antibody.

"Crossing Over": The translocation of genes, genetic material, or part of a chromosome that sometimes occurs at mitosis or meiosis.

Cross-Match: A slang term to describe a compatibility test.

Deletion: Loss of genetic material from a chromosome. Red cells from individuals who have suffered gene deletion are frequently referred to as "deletion cells," e.g., D- -; Jk(a– b–).

Dominant Gene: A gene that is expressed, in terms of antigenic product, when present in double or single dose.

Dosage Effect: The difference in the amount of antigen produced when two like genes are present compared to the amount produced when two dissimilar genes are present. Recognized by the reactions of some antibodies that are dependent on the amount of antigen present for the strength of their reaction.

Gene: The basic unit of inheritance that determines the production or nonproduction of a character. It may be necessary that a series of genes be present in order that a recognizable character (such as a blood group antigen) be produced.

Genotype: A listing of the genes thought, from the results of typing tests, to be present in an individual.

Hemoglobinuria: The presence of hemoglobin or its derivatives in urine.

Hemolysin: An antibody capable of destroying red cells in the presence of complement.

Heterozygous: The state where homologous chromosomes carry nonidentical allelic genes.

Homozygous: The state where homologous chromosomes carry like allelic genes.

Hyperbilirubinemia: Elevated (above normal) levels of bilirubin in serum.

"Immunized" Individual: A person producing an antibody.

Immunocyte: A cell with the ability to synthesize gammaglobulin antibody.

"Incomplete" Antibody: An antibody that attaches to its specific antigen on the red cell surface but that is unable to cause agglutination in a saline system.

Inhibition: Neutralization of an antibody by means of adding the specific substance, with which it reacts, in a soluble form.

In Vitro: Outside the body (usually in a test tube).

In Vivo: Within the body.

Kernicterus: Damage to the brain cells that can be caused by abnormally high levels of serum bilirubin.

Laked: Hemolyzed.

Lectin: An extract made from seeds that has the ability to agglutinate red cells. Lectins commonly used in blood group serology have blood group specificity. Many others are known that agglutinate all red cells.

Linked Genes: Genes located on the same chromosome that are normally inherited together.

Locus (plural Loci): The site of a given gene (or one of its alleles) on the chromosome.

Meiosis: The process of cell division by which the sex cells are formed. For chromosome content of ova and sperm: see *chromosome.*

Mitosis: The process of cell division by which all new cells except sex cells are formed. For chromosome content of cells: see *chromosome.*

Mosaic: As applied to red cell antigens this term is used to describe antigens that are composed of a series of different parts not all of which are always present in all individuals with some expression of the antigens.

Mutation: An alteration in the structure of a chromosome probably at the level of the gene. The mutant form is often recognized in the family of the person in whom the mutation has occurred.

Phenotype: A listing of genes that can be shown to be present by direct typings for antigens on the individual's red cells. Not necessarily a complete listing of all genes present.

Polymorphism: The existence, in a population, of more than one alternative phenotype.

Position Effect: An effect (usually noticed by the amount of antigen produced) exerted on one gene present by other genes simultaneously present on the same or opposite member of a homologous pair of chromosomes. Note: *cis* position involves genes present on the same chromosome as the one whose product is affected. *Trans* position involves genes present on the other chromosome of the pair.

Propositus: The first-studied member of a family. The member whose unusual blood type initiated a study of the blood of related members of the pedigree.

Prozone: A situation where an antibody reacts better when diluted than when used in an undiluted (raw) state.

Recessive Genes: Genes whose presence is detected, directly, only when the genes are present in the homozygous state.

"Sensitized" Cells: Cells coated with, but not agglutinated by, an antibody.

"Sensitized" Individual: A term sometimes used *incorrectly* to describe an immunized individual.

Specificity: The property of an antibody that enables it to combine with one but not other antigens.

Thermal Amplitude: The temperature range over which an antibody is active.

Titer: A measure of the strength of an antibody. Serial dilutions of the antibody are tested against appropriate red cells, and a titer value is assigned based on the highest dilution of the antibody still capable of causing a positive reaction. The titer is correctly expressed as a reciprocal of the serum dilution.

Titer Score: The sum of all positive reactions (graded for strength) from an antibody titration.

Translocation: The transfer of genetic material from one chromosome to another.

APPENDIX 3

Standard Abbreviations for Immunological Terms

Ab	antibody
Ag	antigen
ACD	acid–citrate–dextrose
ACTH	adrenocorticotrophic hormone
ADA	adenosine deaminase
ADCC	antibody-dependent cell-mediated cytotoxicity
AFP	alpha-fetoprotein
AICF	autoimmune complement fixation
AIHA	autoimmune hemolytic anemia
ALG	antilymphocyte globulin
ALS	antilymphocyte serum
AMP	adenosine monophosphate
ANA	antinuclear antibody
ANF	antinuclear factor
BALB/c	inbred strain of mice
BCG	bacillus Calmette-Guérin
BFPR	biological false-positive reaction
BGG	bovine gammaglobulin
BJ	Bence–Jones
BPO	benzyl penicilloyl
BSA	bovine serum albumin
BSS	buffered saline solution
C1, C2, etc.	abbreviations for complement components
cAMP	cyclic adenosine monophosphate
CEA	carcinoembryonic antigen
CF	complement fixation
CFA	complete Freund's adjuvant
CFT	complement fixation test
CGD	chronic granulomatous disease
cCMP	cyclic guanosine monophosphate
CMC	chronic mucocutaneous candidiasis
CMI	cell-mediated immunity
ConA	concanavalin A
CMV	cytomegalovirus(es)
C3PA	C3 proactivator
CSF	cerebrospinal fluid
DEAE	diethylaminoethyl
DNCB	2,4-dinitrochlorobenzene
D–L	Donath–Landsteiner
DNFB	dinitrofluorobenzene
DNP	dinitrophenyl
DTH	delayed-type hypersensitivity
DTP	diphtheria and tetanus toxoid combined with pertussis vaccine
EA	erythrocyte amboceptor (sensitized erythrocytes)
EAC	erythrocyte amboceptor complement
EAE	experimental allergic encephalitis
EBV	Epstein–Barr virus

ECF eosinophil chemotactic factor
ECF-A eosinophil chemotactic factor of anaphylaxis
EDTA ethylenediaminetetra-acetate
ER endoplasmic reticulum
ESR erythrocyte sedimentation rate
F_1 first generation
F_2 second generation
FA fluorescent antibody
Fab antigen-binding fragment
FEV forced expiratory volume in 1 second
FSH follicle-stimulating hormone
GALT gut-associated lymphatic (lymphoid) tissue
GBG glycine-rich beta-glycoprotein
GBM glomerular basement membrane
GCFT gonococcal complement fixation test
Gm allotypic marker on human IgG
GMP guanosine monophosphate
GPA guinea pig albumin
GPC gastric parietal cell
G6PD glucose 6-phosphate dehydrogenase
GVH graft-versus-host (disease)
GVHR graft-versus-host reaction
HCD heavy-chain disease
HCG human chorionic gonadotropin
HDN hemolytic disease of newborn
H&E hemotoxylin and eosin (stain)
HGG human gammaglobulin
HI hemagglutination inhibition
HLA human leukocyte antigen
HMP hexose monophosphate (shunt)
HSA human serum albumin
IEP immunoelectrophoresis
IF intrinsic factor
IF interferon
IFA indirect fluorescent antibody
Ig (A,D,E,G,M) immunoglobulin and other immunoglobulin units and subunits as defined by WHO
Inv allotypic marker on human kappa chain
i.m. intramuscular
i.p. intraperitoneal
ISC immune serum globulin
ITP idiopathic thrombocytopenic purpura
i.v. intravenous
JRA juvenile rheumatoid arthritis
K cells killer cells
KLH long-acting thyroid stimulator
LATS keyhole limpet hemocyanin
LCM lymphocytic choriomeningitis
LD lymphocyte defined
LDH lactate dehydrogenase
LE lupus erythematosus
LIF leukocyte-inhibitory factor
LPS lipopolysaccharide
LT lymphotoxin or lymphocytotoxin
MAF macrophage-activating (arming) factor
MCA methylcholanthrene
MCF macrophage chemotactic factor
2-ME 2-mercaptoethanol
MeBSA methylated bovine serum albumin
MEM minimum essential medium

MER methanol extraction residue (of phenol-treated BCG)
MF mitogenic factor
MHC major histocompatibility complex
MHD minimum hemolytic dilution or dose
MIF migration-inhibitory factor
MLC mixed lymphocyte (leukocyte) culture
MLD minimum lethal dose
MLR mixed lymphocyte (leukocyte) response or reaction
MPO myeloperoxidase
MS multiple sclerosis
MTX methotrexate
MW molecular weight
NBT nitroblue tetrazolium (dye)
NZB New Zealand black (mice)
NZW New Zealand white (mice or rabbits)
OA ovalbumin
OAF osteoclast-activating factor
OT old tuberculin
PA pernicious anemia
PAS periodic acid–Schiff (reaction)
PBS phosphate-buffered saline
PC parietal cell
PEC peritoneal exudate cells
PF/dil permeability factor dilute
PFC plaque-forming cells
PFU plaque-forming units
PHA phytohemagglutinin
PLL poly-L-lysine
PMN polymorphonuclear neutrophils
PNH paroxysmal nocturnal hemoglobinuria
PPD purified protein derivative (tuberculin)
PVP polyvinylpyrrolidone
PWM pokeweed mitogen
RA rheumatoid arthritis
RAST radioallergosorbent test
RBC red blood cell
RDE receptor-destroying enzyme
RES reticuloendothelial system
RF rheumatoid factor
RIST radioimmunosorbent test
SC secretory component
SCID severe combined immunodeficiency disease
SD serologically defined
SK-SD, SKSD streptokinase-streptodornase
SLE systemic lupus erythematosus
SLO streptolysin-O
SLS streptolysin-S
SMA smooth muscle antibody
SMAF specific macrophage arming factor
SNagg serum normal agglutinator
SRBC sheep red blood cells
SRF skin-reactive factor
SRS-A slow-reacting substance of anaphylaxis
SSPE subacute sclerosing panencephalitis
TA transplantation antigens
TCA trichloroacetic acid
TCH tanned cell hemagglutination
TF transfer factor
TFd dialyzable transfer factor
TL thymic-lymphocyte (antigen)
TSA tumor-specific antigen
TSH thyroid-stimulating hormone
TSTA tumor-specific transplantation antigens
VIG vaccinia immune globulin
WBC white blood count

APPENDIX 4

Methods for Detecting Antibodies

Method	Minimal amount of antibody	
(1) Precipitation		
(a) micro-Kjedahl	125	μg
(b) colorimetry (absorbance at 280 nm, Lowry, Nessler)	6–25	μg/ml
(c) optimal proportions	50	μg/ml
(d) ring test	20–30	μg/ml
(e) nephelometry:		
molecular	0.5	μg/ml
bacterial	0.005–0.01	μg/ml
(2) Gel diffusion		
(a) Preer method (double-diffusion tube)	20–50	μg/ml
(b) Oudin method (single-diffusion tube)	12–110	μg/ml
(c) Ouchterlony method (double-diffusion plate)	40	μg/ml
(d) Mancini method (radial-diffusion plate)	10	μg/ml
(3) Immunoelectrophoresis		
(a) standard microtechnique	100	μg/ml
(b) crossed (Laurell technique)	100	μg/ml
(c) radio-technique (with anti-globulin)	0.02	μg/ml

(4) Agglutination		
(a) bacteria:		
Salmonella typhi	0.12	μg/ml
pneumococci	60–120	μg/ml
(b) red blood cell isoagglutinins	0.6–1.2	μg/ml
(c) Rh-positive blood cells–antiglobulin test	0.03	μg/ml
(d) passive hemagglutination with protein-sensitized cells	0.03	μg/ml
(5) Complement fixation (micromethod) with rabbit antibody	1–0.1	μg
(6) Anaphylaxis		
(a) passive cutaneous, with rabbit antibody	0.02	μg
(b) Schultz–Dale test with rabbit antibody	6	μg/ml
(7) Fluorescence polarization	3–10	μg/ml
(8) Fluorescence quenching	40	μg/ml
(9) Radioimmunoassay	μg–ng/ml range	
(10) Specific immunoadsorbent	20	μg/ml
(11) Specific immunoadsorbent-inhibition technique	0.01–0.02	μg/ml
(12) Farr binding technique (ABC-33)	0.05–0.1	μg/ml
(13) P_{80} binding technique	1	μg/ml
(14) Anti-globulin technique	μg–ng/ml range	
(15) Nonprecipitating antibody by coprecipitation	50–100	μg/ml
(16) Diphtheria toxin neutralization (rabbit skin test)	0.01	μg/ml
(17) Specific bioassay	μg-ng/ml range	
(a) enzyme activity		
(b) count of viable bacteria		
(c) syphilis flocculation test		

(18) Neutralization tests (a) phage (ϕX-174) (b) viruses (polio etc.)	1 molecule of antibody* μg
(19) Fluorescent dye uptake	1 molecule of antibody*†
(20) Cytotoxic antibody technique	1 cell coated with antibody*
(21) Jerne plaque technique	1 antibody-producing cell*
(22) Immunocytoadherence (a) red cells (Rosette formation) (b) bacterial adherence	1 cell coated with antibody*
(23) Coons tissue fluorescence technique	1 antibody-producing cell*
(24) Ethidium bromide uptake	1 molecule of antibody*
(25) Cunningham monolayer technique	1 antibody-producing cell*

* This is a theoretical limit of detection. Technical problems, such as sampling, greatly reduce the sensitivity of the method.

† Applicability with β-galactosidase only as the antigen.

INDEX

Index